Range Rover

The Complete Story

James Taylor and Nick Dimbleby

CROWOOD
AutoClassic

First published in 1995 by
The Crowood Press Ltd
Ramsbury, Marlborough
Wiltshire SN8 2HR

Britsh Library Cataloguing-in-Publication Data
A catalogue record for this book is available from the British Library

ISBN 1 85223 910 7

Acknowledgements
We were very fortunate in receiving a great deal of enthusiastic help, and we are more than pleased to acknowledge those who provided time, vehicles, knowledge or photographs. In alphabetical order, we would like to give our sincere thanks to: Jack Baines, Graham Bannock, The late Gordon Bashford, Roger Crathorne, British Motor Industry Heritage Trust (especially Anders Clausager), Caffyns of Lewes, Vince Cobley, Dakar Cars (Barry Chantler), The Dunsfold Land Rover Trust (especially Brian and Philip Bashall), Dutton Forshaw of Maidstone, Steve Dyer, ETT (especially Christian Eggenspiller), Glenfrome of Bristol (especially Kenneth Evans), Keith Hales, Peter Hobson, *International Off-Roader and 4 × 4 Owner* magazine (Colin Dawson), Spen King, The Land Rover Centre, Huddersfield (Mark Griffiths), Land Rover Ltd (especially Nick Argent and Vin Hammersley in the Press Office, and Paul O'Connor in Photographic), Land Rover North America (especially Bill Baker and the press team), *Land Rover Owner* magazine, The Metropolitan Police, Geof Miller, Bob Morrison (of Military Scene), Overfinch Ltd (Douglas Cooke and Neil Underwood), Tony Poole, Gary and Craig Pusey, Rover France (especially Jean-Pierre Lafitte and Steve Norman), A E Smith of Kettering (John Perkins), Taunton Land Rover (especially Rod Warren), Vantagefield Ltd (Graham Chalkley and David Linder), Warn Industries.

Picture Credits
All photographs supplied by the authors, Land Rover Ltd, Dutton Forshaw of Maidstone, Glenfrome of Bristol, Land Rover North America, Bob Morrison of Military Scene, Overfinch Ltd, Tony Poole, A E Smith of Kettering and Vantagefield Ltd

Printed and bound by The Bath Press

Contents

Range Rover Evolution

June 1970	Range Rover introduced
January 1973	First major revisions, including PAS option
October 1975	First Option Pack to improve equipment levels
September 1979	Facelift, including black bumpers and new badges
March 1980	Limited volume Monteverdi four-door conversion
February 1981	Two-door Range Rover In Vogue special edition model
July 1981	Four-door production model introduced
August 1982	Three-speed automatic option introduced
July 1983	Five-speed manual gearbox replaced four-speed
June 1984	Range Rover Vogue introduced as regular production model; new instruments and interior trim
October 1985	Fuel injection and front spoiler for Vogue models only; new four-speed automatic
April 1986	Range Rover Turbo D introduced with 2.4-litre VM engine
November 1986	Specification upgrade with horizontal-bar grille and concealed fuel filler flap; all four-door models now fuel-injected
	Two-door models discontinued for UK market
March 1987	Range Rover launched on US market
March 1988	Vogue SE introduced
October 1988	Chain-driven transfer box replaced gear-driven type; 3.9-litre engine for USA only
October 1989	3.9-litre engine replaced 3.5-litre for all markets; 2.5-litre VM diesel replaced 2.4-litre in Turbo D; ABS introduced
September 1990	Anti-roll bars introduced (on limited-edition Range Rover CSK)
October 1992	Long-wheelbase Vogue LSE introduced; air suspension for top models; Land Rover 200 Tdi engine replaced VM 2.5-litre
January 1994	Two-door production ended
March 1994	New dashboard with airbags; new R380 five-speed gearbox; 300 Tdi engine replaced 200 Tdi; SE trim and automatic options on diesels
October 1994	Range Rover renamed Range Rover Classic on the introduction of the second-generation Range Rover
June 1995	Range Rover celebrated 25 years of continuous production

1 In the Beginning

The origins of the vehicle which became the world-beating Range Rover can be traced back more than thirty years to the mid-1960s. At that time, the Rover Company, makers of luxury saloons, executive saloons and Land Rovers, had just established a Market Research department for the first time, and the man in charge of it was a young economist called Graham Bannock.

> I was doing some pure desk research, which established that if you looked at the statistics on the sales of Land Rovers and sales of competitive vehicles ... and if you broke those down into class of customer, it was the so-called leisure users that were growing faster. Military users were fairly static, agriculture and so on was also fairly static. The real growth was coming from people who were buying Land Rovers to tow caravans, to go on holiday, from architects and surveyors who have to go across rough country ... from people who were living in suburban houses but wanted to project the image which most 4x4 buyers today want to project, of being real country types.
>
> It was a very clear picture. This was worldwide, particularly in the US, but just everywhere. Even in the Middle East, which was a big market for Rover at the time, sheikhs were buying them for fun. It was that gave me the idea that a vehicle which combined the best qualities of the Rover car and the Land Rover would find a big market, even though such a vehicle would inevitably cost more than a Land Rover.

Over the summer of 1965, Bannock visited the USA, where Rover were struggling to sell their saloons and four-wheel drives. One purpose of his visit was to see how the company might improve its sales in that market, but he was also keen to test out his theories.

What he saw confirmed that he had been right: one of the most striking automotive success stories of recent years had been that of the International Scout, a four-wheel drive recreational and leisure vehicle introduced in 1961. The Jeep Utility Wagon, another 1961 arrival, was also selling strongly; and during 1965, Ford introduced their competitor in the shape of the Bronco. Chevrolet were already rumoured to be working on their equivalent, which materialized as the Blazer a year later.

It seemed to Bannock that Rover could and should produce such a vehicle, using the four-wheel drive expertise they had gained from two and a half decades of making Land Rovers. He came up with an outline description of what he had in mind, and on his return to the company's Solihull headquarters in August 1965 he shared his ideas with Spen King. King was then in charge of Rover's New Vehicle Projects department, which had been set up specifically to look at future projects and ideas which might go into them, and he already had a formidable reputation as an engineer from his work on gas turbine cars in the 1950s and later on the Rover 2000 saloon project.

King's right-hand-man and chief layout designer was Gordon Bashford, who had then been with Rover for more than thirty years and had been responsible for the basic chassis design of every vehicle the company had built since 1945, saloons and Land Rovers alike. The two of them set to work on Bannock's idea, and by the early months of 1966 they had come up with some paper schemes for a vehicle to suit the market he had identified. At this stage, these schemes were purely conjectural: no budget had been granted for such a vehicle to be designed, and King and Bashford were simply toying with ideas. Nevertheless, they did arrange for Rover's North American subsidiary to send over examples of the International Scout, Ford Bronco and Jeep Wagoneer for examination.

The vehicle King and Bashford had in mind was very clearly a station wagon type of Land Rover, but it had one very significant difference from the Land Rovers of the time. Where these had leaf-spring suspension, the King-Bashford proposals incorporated long-travel coil springs. This would have been anathema to the Land Rover side of the company if they had known about it: orthodox thinking then was that Land Rovers needed the harsh ride provided by leaf springs so that their drivers would be discouraged from pushing them too hard in rough terrain and so wrecking the vehicles! However, King was sure of the rightness of his idea. He believed that long-travel springs would not only give the softer ride needed in a road vehicle, but would also give more of the axle articulation needed for extreme off-road work. To be certain, he had tried a Rover 2000 saloon (which had long-travel coil springs) off the road, and had come away convinced.

These early paper schemes also incorporated the 2,996cc straight-six Rover engine which then developed 134bhp in the big Rover 3-litre saloon and was under consideration in detuned 110bhp form for Land Rovers. However, by this stage Rover already had the manufacturing rights to the General Motors Buick 3,528cc light-alloy V8 engine, and before long King and Bashford incorporated this engine into their schemes. It brought with it many advantages, being shorter and lighter than the 3-litre six, offering a lot more power and torque, and having the V8 configuration which would have a strong appeal in the USA.

In the meantime, Graham Bannock had been working on a detailed report which fleshed out his idea of a recreational Land Rover vehicle. This was completed on 19 July 1966 and when he showed a copy to King and Bashford, he was amazed to discover that they had taken his original ideas so seriously and that the Land Rover they were drawing up so exactly fitted the bill. 'It *was* the vehicle I was describing,' remembers Bannock.

THE INITIAL SPECIFICATION

By the time of Bannock's report, the paper schemes were quite advanced. Like existing Land Rovers, the new vehicle was to have a tough ladder-frame chassis with box-section side members. Land Rover practice would also be followed in the transmission, which was to be a selectable four-wheel drive type. There would be beam axles like a Land Rover, too, but of course these would have coil springs and locating arms instead of the Land Rover's crude semi-elliptic leaf springs. In addition, Spen King wanted disc brakes to cope with the anticipated high performance; although experimental Land Rovers had been built with disc brakes, the production types of the mid-1960s still had drum brakes all round.

Body design had also been given careful consideration, and once again the Rover 2000 provided the necessary inspiration. This was constructed around a rigid 'base-unit', an inner skeleton to which unstressed outer panels were bolted and which recent tests had shown to be extremely strong in a collision. As the new Land Rover was being designed with the safety-oriented US market firmly in mind, it was vital to use a crash-resistant body structure. So Gordon Bashford designed its body as a steel inner skeleton to which could be bolted unstressed outer panels of the same corrosion-resistant alloy as was used for Land Rover bodies. With this body bolted to the rigid box-section chassis, the result would be an enormously strong structure.

Bashford's design had also taken account of the likelihood that the new vehicle would be shipped out in CKD (completely knocked down) kit form for assembly in some overseas territories. Thus, the bodyframe was designed with flat sections which bolted together like a self-assembly garden shed. There was a single drawback – although it did not appear to be one at the time. It was going to be easier to get the necessary strength in a two-door body than in a four-door type, and so Bashford had designed his body with just two doors. As the American recreational vehicles Rover had studied also had two-door bodies, Bannock, King and Bashford all believed that this was what the market would want. At a later stage in development – too late for any major changes to be made – it became clear that this had been a wrong assumption.

THE 100-INCH STATION WAGON

Although Spen King, Gordon Bashford and Graham Bannock were convinced of the potential for the recreational Land Rover, Technical Director Peter Wilks was not so sure. Nevertheless, he did agree to allocate a budget to the project, and more or less the next thing to happen was that a Land Rover engineer, Geof Miller, was seconded to New Vehicle Projects in July 1966 to provide the four-wheel drive knowledge which King's department did not have. Among his first tasks were thorough engineering and test analyses of the Scout and Bronco, and a briefer analysis of the Wagoneer; in these, he was assisted by Nick Wilks, another Land Rover engineer who was temporarily lent to King's department.

That summer also saw Miller doing some calculations which convinced him that the plan to use standard Land Rover axles in the new vehicle was likely to lead to trouble. The problem was that the V8 engine delivered much more torque than any previous Land Rover power unit, and when that torque was all put through the rear axle – as it would be in two-wheel drive – the half-shafts and differential would be put under unacceptable strain. One way round the problem was to use a heavy-duty axle, but Spen King was strongly opposed to this because its extra weight would spoil the ride quality he was aiming for. So Miller suggested using permanent four-wheel drive, an idea Rover had toyed with back in the 1950s. This would enable the torque to be split between two relatively light axles and would thus not compromise ride comfort.

King liked the idea: he had incorporated permanent four-wheel drive in the experimental Rover T3 gas turbine car some ten years ealier, and he knew that it improved road-holding and minimized the risk of wheelspin under hard acceleration. He was finally convinced after some comparison trials in the snow around Edge Hill in the early months of 1967. For these, Geof

Miller supplied an experimental V8-powered 88-inch Land Rover and a similar vehicle equipped with a permanent four-wheel drive transmission which had lain in the Rover Experimental Stores since the 1950s. King subsequently asked the Transmissions Department to design a new transmission incorporating permanent four-wheel drive and the centre differential necessary to allow front and rear axles to rotate at different speeds.

As King and Bashford were both primarily car men, they had meanwhile been finding out more about the demands placed on four-wheel drive vehicles in off-road driving. In October and November 1966, Geof Miller organized two trips down to Land Rover's off-road test centre at Eastnor Castle in Herefordshire. A motley collection of vehicles went on these trips, including standard production Land Rovers, experimental Land Rovers, and seven competitive vehicles: a Nissan Patrol, a Toyota Land Cruiser, a Steyr-Puch Haflinger, a Farmobil, an Austin Gipsy and New Vehicle Projects' own International Scout and Ford Bronco.

Many years later, King said he doubted whether these sessions had taught New Vehicle Projects very much of direct relevance to the design of the new Land Rover. However, Geof Miller's recollection is that most participants said at the debriefing how useful the exercise had been. He also believes the Eastnor sessions convinced everyone that the 3.5-litre V8 (which was in an experimental 88-inch Land Rover) would be the only worthwhile engine for the new vehicle.

By now, Gordon Bashford had started work on the overall vehicle layout. Towards the end of the year, he had completed what looked like a reasonably definitive package, in which the wheelbase worked out at 99.9in. 'So I suggested we should round it up to 100 inches,' he recalled some twenty years later, 'and we called it the Land Rover 100-inch Station Wagon.' Things were far from completely settled yet, however. Bashford continued to refine his design, and the early part of 1967 saw him looking at leaf springs instead of coil springs to simplify axle location. While these would undoubtedly have suited the more traditional elements on the Land Rover engineering staff, they fortunately went no further than the drawing-board!

In addition, the early part of 1967 saw a proposal to save costs and widen the vehicle's appeal by having a two-model range, the cheaper version with a four-cylinder engine and the more expensive with the V8. Initially, the four-cylinder was expected to be the 90bhp engine from the Rover 2000, but the focus later switched to the 2.25-litre Land Rover engine. And although no four-cylinder prototype was ever built, the 'economy model' remained an element in management policy discussions until it was abandoned at the end of 1968.

Bashford's layout drawings were quickly turned into a full-size buck in the Mock-Up Shop, and this was examined by Rover's senior management in January 1967. King also wanted David Bache's Styling Department to see what they could do with the basic outline, but Bache told him they were too busy to help in the short term. So King and Bashford sketched the vehicle up themselves, after prevailing upon Bache's assistant Geoff Crompton to give them a little freelance advice.

Work started on building a first prototype vehicle in the Jig Shop in the early spring of 1967, and at the same time Graham Bannock drew up a market research questionnaire so that Rover could check their ideas really were on track. This was passed on to an agency which conducted the actual survey in Britain during the early summer of 1967. For the purposes of the survey the new vehicle was known as Concept Oyster, and interviewees were shown an early concept sketch in order to gauge its size and general appearance. Reactions were very favourable. Bannock recalls that the research was carried out in two stages and that the interviewees were told only in the second stage that the vehicle would be a Rover. It is an interesting reflection on the company's standing at the time that the association with the Rover name produced far more positive reactions.

The information given to Oyster interviewees sheds further light on the way Rover saw their new vehicle at this stage in its development.

> Oyster is not a converted passenger car but a purpose-built estate car with a unique combination of features. It will offer safe, fast and relaxing travel on motorways, on the road, across fields, sands, snow or ice. It is easy to drive and handle, will tow boats or caravans, carry milk churns, be used as a camping vehicle or you can use it to go shopping or to go to the opera. Many extras from roof-racks to winches to air-conditioning will permit the owner to adapt the vehicle for his own special requirements.

The interviewees were also told that there would be a three-speed automatic transmission alternative to the standard four-speed. Obviously, some of this was not so much reality as intention, as not a single vehicle had been built when Graham Bannock wrote those words!

Nevertheless, the first prototype was up and running by July. Bannock remembers it very clearly:

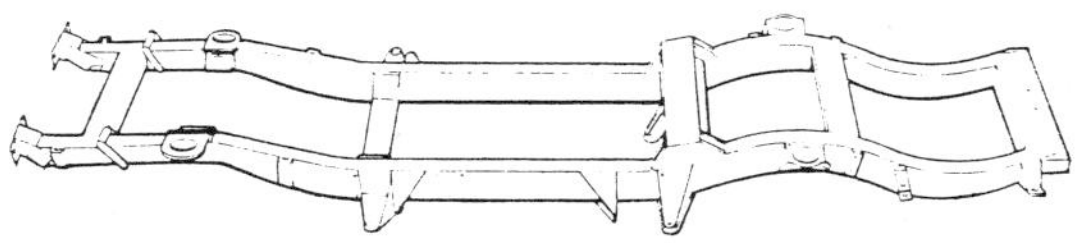

The basis of the Range Rover was this immensely strong box-section chassis-frame, designed along similar principles to the Land Rover chassis.

This is almost certainly the first picture ever taken of the completed Range Rover prototype on the day it first ran in July 1967. At the wheel is probably Peter Wilks, Spen King is sitting beside him on the wooden box which then did duty as a passenger seat, and the picture was taken by Graham Bannock.

Additional strength came from the self-supporting body frame, which formed the inner panels of the vehicle.

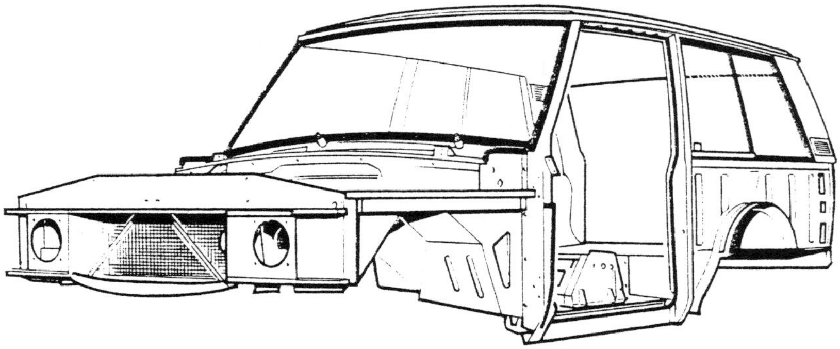

> Spen King rang me up and said, 'Look, we've just got the prototype running. Do you want to have a spin?' So I rushed out of my little office, and Spen and Peter [Wilks] were waiting down by the Engineering Block, and we drove off and went round, and I took my camera with me and took some photographs.

At this stage, the prototype had only a driver's seat and any passengers had to sit on a wooden packing case on the floor, but the incident well reflects the level of enthusiasm which New Vehicle Projects were putting into the 100-inch Station Wagon. On the Land Rover side of the house, by contrast, there was still a good deal of scepticism, and Geof Miller recalls that most Land Rover people believed the 100-inch Station Wagon was doomed to failure.

This first prototype was a very rough-and-ready vehicle. Although it had a specially-constructed box section chassis with coil spring suspension, its transmission, axles and brakes were all off-the-shelf Land Rover items, and the rest of the vehicle had been cobbled together. Not surprisingly, a number of problems soon became apparent. With the vehicle unladen, it was only too easy to spin the rear wheels in almost any gear. The drum brakes, albeit supplemented by a servo, also provided some interesting experiences in a vehicle which Research Department engineer Keith Hales remembers as capable of an indicated 110mph (180km/h).

The front suspension was also unsatisfactory. To save time, the Rover engineers had copied its radius rods from the Ford Bronco and located the axle with a diagonal link system. However, the result was excessive steering kickback over all types of terrain: 'It was just awful!' remembers King. However, the basic structure which Gordon Bashford had proposed, with unstressed skin panels on a stressed-steel body frame, did prove satisfactory in the pavé tests to which the vehicle was subjected over the next few months.

It also went out on the road, when it was thinly disguised with large enamel 'Velar' badges front and rear. Although Gordon Bashford was convinced for many years that this name stood for Vee-Eight LAnd Rover, the reality was rather more prosaic: the name had been used earlier on a prototype Rover sports car which was intended to wear Alvis badges. The Velar name, though not the enamel badges, would also be used to disguise later prototypes when they were out on public roads.

This prototype was crucial in securing the support of Technical Director Peter Wilks. After trying it out, 'he came round very quickly and gave it his full support', remembers Graham Bannock. He encouraged King to write a formal Product Proposal, and he and Managing Director William Martin-Hurst convinced the Rover Board to grant their approval in July 1967. Wilks went on to be 'enormously supportive and very creative' in the subsequent stages of the project, according to Graham Bannock.

Now committed to the 100-inch Station Wagon, Rover began to bring in other departments such as Planning, whose task was to decide which components could be made in-house and which would have to be sourced elsewhere. A fully detailed programme of work was drawn up, and was agreed by the Main Board in January 1968. All departments involved now knew what they had to do and when they had to do it in order for production of the new vehicle to start in December 1969 and enough examples to be ready for sales to begin in February 1970. There would of course be some slippage in what was acknowledged to be a tight schedule – but Solihull did stick remarkably closely to this programme.

David Bache's first styling proposals were rejected as impractical, although they would certainly have made the vehicle distinctive. This quarter-scale model was made at the end of June 1967.

STYLING

In the meantime, David Bache's department had found some time to work on the vehicle. Beginning in June 1967, they produced some scale model proposals which initially gave the 100-inch Station Wagon a more streamlined, car-like appearance. Bache himself explained several years later that he had been trying to get away from the boxy shape of the King–Bashford prototype. However, Spen King pointed out that these stylish shapes would not be suitable for the knock-down overseas assembly which was already in the back of the designers' minds, and so Bache took a second look.

This time, he decided simply to tidy up the basic shape of the existing prototype. The first quarter-scale model of a 'tidied' 100-inch Station Wagon was done in July, and by September all parties were in agreement that this was the way to go. A full-size model was therefore built, and Bache badged it as a Road-Rover, borrowing the name from an aborted project of the 1950s. However, no decision had actually been taken about a name for the new vehicle; officially, it was still simply a Land Rover 100-inch Station Wagon.

Although David Bache's final styling proposal was very similar in many ways to the King-Bashford design executed on the first prototype, it is important to recognize just what a difference his involvement made. At the front, he provided a strong

By the autumn of 1967, all parties had agreed that the best way forward was simply to tidy up the body shape used for the first prototype. This full-size wood-and-clay buck was built in the Styling Department and was badged, for want of anything better, as a Road-Rover. A lot of work went into giving the vehicle a distinctive 'face' and two alternative proposals are shown here. The style eventually chosen combined elements of both proposals.

visual identity with a bold black grille, large rectangular sidelights and indicator lamps, and a rugged-looking bumper. He gave greater definition to the bonnet shape with the castellated features at the leading edges (a feature which also suited the production engineers as it gave valuable additional strength to the bonnet panel). A quarter of a century later, market research showed that the public considered those castellations an essential part of the Range Rover 'face', and the team styling the second-generation model had no choice but to incorporate them in their own vehicle.

Bache's restyle also gave the sides and rear of the vehicle much more character. Where the King–Bashford prototype had twin swage lines running down its sides to give stiffness to the panels, Bache turned them into a side indentation feature. He matched the chunky front lights by large wrap-around lamp clusters at the rear, added another rugged-looking bumper, and brought the side indentation feature round to give more character to the lower tailgate. The result was a striking-looking vehicle with a much more attractive appearance than any existing four-wheel drive estate, and yet one which also reflected the ruggedness which was an essential part of its character.

King and Bashford had of course not even attempted to style the interior of their prototype. They had simply had some very basic Land Rover-style seats made up in order to get the vehicle into running condition. So the Styling Department had a completely free hand in this area. Bache entrusted the job to a team under Tony Poole, and work began in the spring of 1968.

The stylists proceeded on the assumption that there would be two levels of trim – standard for the four-cylinder model and de luxe for the V8 – and laid out their timetable accordingly. As a first stage, they created a sculpted facia shape which would be simple to manufacture and would suit

The full-size buck incorporated letter-box door handles, a sculpted air outlet on the quarter-pillar and a cover over the fuel filler. All these features disappeared before production began, but the fuel filler cover did return – nearly twenty years later! Those wheel trims are from a Rover 2000.

A separate full-size interior buck was created in the Styling Department. Note the ingenious release catches for the passenger seat (the driver's seat had a different arrangement) and the rather unsatisfactory saftey belt installation. This interior would have been rather plusher than the one which eventually went into production: note the console panel on the transmission tunnel and the wraparound shoulders of the rear seat, which also appears to have the one-third / two-thirds split feature not introduced in production until the mid-1980s.

both right-hand and left-hand drive vehicles. They then moved on to the door trims, which rapidly took on their definitive shape. Finally, it was the turn of the seats, where there was a particular problem because the front seats had to be arranged to tip forwards and allow access to the rear.

Safety was a major consideration for the 100-inch Station Wagon because Rover hoped to sell it in the increasingly safety-conscious US market. As a result, the vehicle would have to be equipped with front seat safety belts at the very least. However, mounting the belts to the B-posts would have put them in the way of rear seat passengers trying to get in or out, while mounting them to the floor so that they ran up and over the seat backs would have caused difficulties when the seat backs were tipped forwards.

So Rover decided to mount the belts to the seats themselves and to mount the seat bases very firmly to the floor. The stylists had to work in tandem with safety engineers to reach a solution which would also allow the seats to slide back and forth and to tip forwards, but in due course they achieved their aim. These seats were then unique in the automotive world and Rover had to seek special agreement for their installation from the legislative authorities in many of the territories where the new vehicle was to be sold.

However, to meet the final deadline, the stylists had to truncate their programme somewhat. Although they had started work on the de luxe interior, they were eventually asked to come up at short notice with a production-ready interior which could be made at a relatively low cost. So the de luxe seats and the optional rearward-facing bench with which they had been experimenting were sidelined, and, in Tony Poole's words, 'it was all hands on deck to get the interior ready in time! So we never did get to do our luxury interior and the original Range Rover was always a lot more spartan than it was ever intended to be.'

THE SECOND PROTOTYPE

Although the production body styling was pretty well established by the end of 1967, the stylists had to review some items which the production engineers thought would be impossible to mass-produce. As a result, the production body design (which meant both styling and structure) was not approved until January 1968. In the mean time, the Project Team needed a second prototype to test improvements in the chassis design, and so they built a vehicle which was very similar in its general appearance to the first prototype. This one was started in the early months of 1968 and was registered for road use in April or May.

The second prototype had left-hand drive, and it once again had selectable four-wheel drive because the new transmission Spen King had asked for was not yet ready. However, it also incorporated disc brakes and a Boge Hydromat self-levelling strut at the rear. This kept the back end at its normal ride height even when heavily laden, and so allowed softer springs to be fitted to improve ride quality. Prototype number two also had a modified front suspension, which had been drawn up by Joe Brown, Gordon Bashford's assistant, and used a Panhard rod to control the lateral movement of the front axle which had caused the steering kickback on the first prototype. Tests showed that the redesign had been successful in all areas: 'This time,' says Spen King, 'we knew we had got it right.'

However, this second prototype was the last one with which Spen King would be directly involved. One result of the amalgamation of Leyland (Rover's owners since 1967) with British Motor Holdings early in 1968 was a management reshuffle. King was summoned before Sir Donald Stokes, Chairman of what was now the British Leyland Motor Corporation, and was told he was being promoted and given overall responsibility for Triumph Engineering. 'I was flabbergasted,' remembers King, 'but I wasn't given any choice in the matter.'

So from June 1968, Gordon Bashford took charge of New Vehicle Projects, and with it he assumed overall responsibility for the 100-inch Station Wagon. In practice, however, Bashford was heavily occupied with other work, and as the main design phase was over, it was left to Geof Miller to

The engine chosen for the Range Rover was a low-compression version of the 3.5-litre light-alloy V8 used in Rover saloons. This picture shows it mated to the LT95 four-speed gearbox and integral two-speed transfer box.

keep the project on course and to report directly to Peter Wilks. As the workload on the 100-inch Station Wagon had increased, so the project team had grown, and by this stage Miller had been joined by two more Land Rover engineers: Nick Wilks had been replaced by Roger Crathorne during 1967, and Alan Wood had followed.

100-INCH BECOMES RANGE ROVER

November 1968 was the next watershed in the development of the vehicle. That month, Geof Miller issued a Technical Brochure to all departments involved in the project, in which he resuméd the specification of the 100-inch Station Wagon. By this stage, it was clear that the V8 engine would be used in all the first vehicles, but that a four-cylinder model with the 2.25-litre Land Rover engine would be ready later. However, the automatic transmission was no longer under consideration; all vehicles would have the four-speed manual gearbox. November also saw some thought being given to a name for the vehicle, and it was probably during this month that stylist Tony Poole put forward his idea. 'We were playing with names like Land Rover Ranger,' he remembers, 'and I said, why not call it a Range Rover? Everybody seemed to think that was a good idea!'

The name was formally adopted at a meeting on 18 December when the Project Team met representatives of the Rover Sales Department to discuss a number of issues. At this same meeting, the plan for a four-cylinder model was scrapped (much to

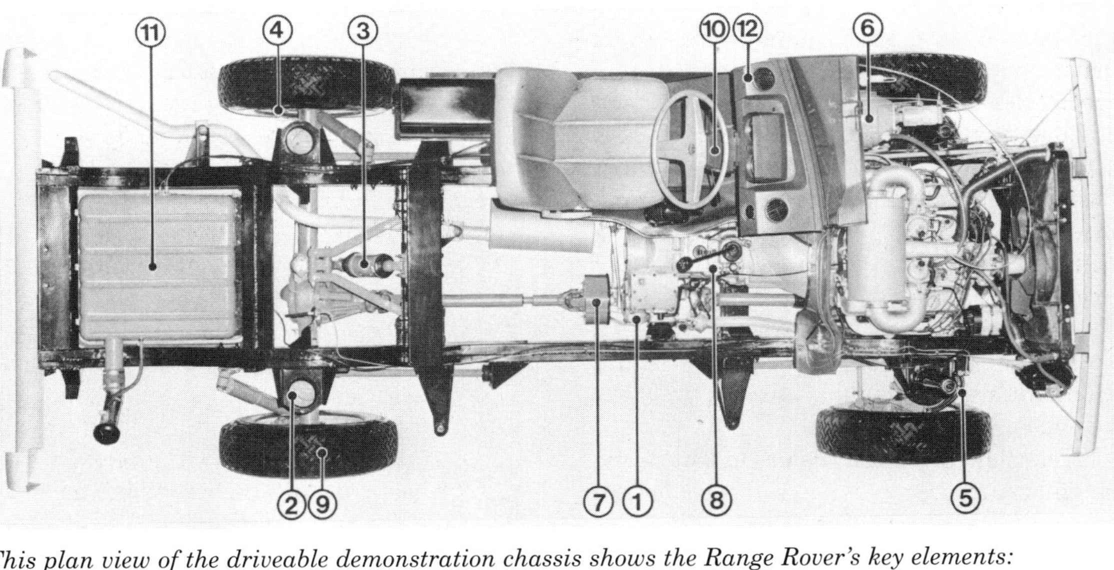

This plan view of the driveable demonstration chassis shows the Range Rover's key elements: 1. Centre differential; 2. Coil springs on beam axles; 3. Boge self-levelling unit; 4. Disc brakes on all four wheels; 5. Dual-circuit front brakes; 6. Brake servo; 7. Drum-type parking brake on the transmission; 8. Four-speed primary gearbox; 9. Radial-ply tyres; 10. Collapsible 'safety' steering column; 11. 19-gallon (86-litre) fuel tank; 12. Impact-absorbing facia. Note also how the rear dampers are 'staggered', with one facing forwards and the other backwards. This was done to prevent axle tramp.

the relief of Peter Wilks, whose department was already overstretched) when Sales said they did not think it would help them to sell any extra vehicles. Sales also went on record as saying that the Range Rover's two-door configuration was 'a tragedy', and that they were convinced there would be customer pressure for a four-door model. In that they were right; in their belief that the Range Rover would have snob appeal they were also right; in their estimates of annual sales between 12,500 and 13,500 they were almost right; but their prediction of an eight-year production run for the new model was to prove a serious underestimate.

FINAL DEVELOPMENT

There were now just twelve months to the scheduled start of production in December 1969. As yet, no prototype with the production body styling or permanent four-wheel drive existed, and by the time one had been completed in April 1969 it must have been obvious that time was running out. This prototype – the third in the series – was the first one to have only two front seats with integral safety belts, the first to have the Rostyle pressed steel wheels which had existed as mock-ups as long ago as February 1968, and the first to have the new high-speed dual-purpose tyres which Michelin had developed specially for the vehicle.

This third vehicle was the final Engineering prototype. The three vehicles which followed in quick succession between August and October 1969 were test prototypes, and two of them were also used for styling trials of details such as badging. By this stage, Land Rover Engineering chief Tom Barton was insisting that the vehicle should bear some form of badging to link it to existing four-wheel drives from Solihull (even though he had not been keen on the vehicle in the beginning), and so an oval 'by Land-Rover' badge was designed for the tailgate. The fourth prototype was used for development work until it became a wheel test rig in the later 1970s, and was eventually scrapped. Prototypes five and six went out to Algeria in November 1969 for a 3,500-mile (5,600km) hot-weather endurance test, and prototype five was then subjected to the barrier-crash test at MIRA. In December, a final complete prototype was assembled to test production build methods and an eighth prototype chassis was built for use as a driveable demonstration unit. And, at the end of December 1969, the first three pre-production Range Rovers came off the newly-installed production lines in the South Block at Solihull.

THE START OF PRODUCTION

Production proper was still several months away, however, and it would be some time before Range Rovers were available in any quantity. The idea of announcing the vehicle at the Geneva Motor Show in March 1970 which had long been in the air at Solihull could not now become reality. So Rover started to think of a launch in the summer. Ambitious plans to take the press on a mini-safari in North Africa were abandoned, and the company settled instead on a quieter affair in Cornwall; after all, Britain would be the only market where the Range Rover would be available for the first six months of its life, so that the new vehicles would not be far from Solihull if they developed teething troubles.

Teething troubles certainly did show up on the production lines in the early months of 1970. One Land Rover dealer who was allowed to see the new vehicles being built remembers that the assembly workers were having to work out some things as they went along – such as where to drill a hole in the bulkhead so that the choke cable could pass through it! Manufacturing problems also caused the aluminium-alloy bonnet to be changed for a steel one after several vehicles had been built: the problem was that the alloy would not 'draw' very well into the castellation features at the panel's leading corners, and the bonnets tended to split in this area.

The facia panel was also changed early on. The first batch of thirty had been run off with a smooth finish in order to minimize tooling costs; once Rover were satisfied that the panels fitted correctly, the tools were modified to give a grained finish, and all subsequent facia panels were of this type. Early assembly trials also showed that the passenger side of the facia tended to droop, and so a reinforcing bracket had to be added. Then production was simplified by fitting silver-coloured wheels with all six body colours; initially, wheels enamelled in Sahara Dust had been specified with Masai Red, Lincoln Green and Sahara Dust itself.

Most of these changes had been made by March or April, when the first of a batch of twenty vehicles intended for use on the press launch in June were built. The early ones may initially have been built with alloy bonnets, but all of them had the grained facia and by the time of the launch they all conformed to production standard. Once they had been built, the production lines started turning out vehicles for the start of showroom sales on 1 September.

The press launch was based at the Meudon Hotel, between Falmouth and the Helford River, and those who attended were able to enjoy driving the twenty specially-prepared vehicles in some delightful scenery. For more serious work, they went on to Blue Hills Mine, an area of steep and rough ground near St Agnes which was better known as the site of the Motor Cycling Club's classic Land's End trial. Here, the Range Rovers showed that they were every bit as capable on rough terrain as a rough-riding utility Land Rover. It was a convincing demonstration of the vehicle's astonishing combination of Rover saloon qualities with those of the Land Rover.

The Engineering Prototypes

The first six prototypes were used for pre-production development and testing. The seventh vehicle was built to test assembly methods. The eighth prototype was a driveable demonstration chassis, and spent many years in the Service Training School at Solihull, gradually being modified to keep pace with the production specification.

The first, fourth and sixth prototypes had right-hand drive. All the others, including the demonstration chassis, had left-hand drive.

Number	*Registration*	*Built*	*Fate*
LR100/1	SYE 157 F	May–July 1967	Scrapped
LR100/2	ULH 696 F	January 1968	Scrapped
LR100/3	AGN 316 G	April 1969 approx.	Not known
LR100/4	AMV 287 H	August 1969 approx.	Scrapped
LR100/5	WYK 315 H	October 1969 approx.	Scrapped
LR100/6	AOY 289 H	September 1969 approx.	Survives
LR100/7	YVB 150 H	January 1970	Survives
LR100/8	Unregistered (Chassis only)	December 1969 approx.	Survives

Development never really stopped: straight after the press launch in June 1970, NXC 239 H was driven into the Styling Studios at Solihull, where it was used for detail mock-up work.

The Pre-Production Prototypes

A very high proportion of the pre-production prototypes still survive twenty-five years after they were built, although not all are in roadworthy condition. The first three of these vehicles were built in December 1969, and the others were assembled in the early months of 1970. Many of them originally wore Velar badges to disguise their identity when they were out on the roads before the public launch of the Range Rover in June 1970.

There were twenty-five right-hand drive examples (in the 355 chassis number series) and just three with left-hand drive (in the 358 series). Several joined the Engineering Fleet and were given a Fleet Number (for example 35500001A became 100/9); others went to the Quality Control or Service Departments; and three (35500016A, 35500021A and 35500022A) were used for Management Assessment of the new product, acquiring fleet numbers MA1 to MA3 respectively.

Two examples (35500010A and 35500025A) were converted to 110-inch (2,794mm) wheelbase for prototype work, and a third (35500002A) was converted to a six-wheel chassis for the prototype Carmichael Commando fire tender.

Surviving records suggest that chassis number one (35500001A) was the first Range Rover to go down the assembly line – although it was quite common for vehicles not to go down the line in strict chassis-number order. From the assembly line, it passed to the Despatch Department, where it was joined by the second and third vehicles. However, the first vehicle to leave the Despatch Department was 35500003A. It seems that the Rover Publicity Department wanted a blue vehicle for photographs, and that 35500003A was used because it was then the only blue Range Rover in existence! The three left-hand drive pre-production vehicles were not built until the end of March 1970, and records suggest that the first of these to go down the lines was actually chassis number three.

All these vehicles were registered in the YVB ... H series, except for 35500006A and (possibly) 35800003A. However, there was no direct correlation between chassis number and registration number.

All vehicles were originally painted in one of the standard production colours except for 35500001A and 35500004A, which had experimental finishes not adopted for production.

Chassis	*Registration*	*Colour*	*Fate*
(RHD)			
35500001A	YVB 151 H	Deep Olive	Survives
35500002A	YVB 152 H	Lincoln Green, later Red	Converted to 6x4 by Carmichael, survives
35500003A	YVB 153 H	Tuscan Blue	Survives
35500004A	YVB 154 H	Amazon Green	Survives
35500005A	YVB 155 H	Davos White	Survives
35500006A	UXC 159 J	Bahama Gold	Survives
35500007A	YVB 177 H	Masai Red	Survives
35500008A	YVB 160 H	Masai Red	Survives
35500009A	YVB 157 H	Davos White	Survives
35500010A	YVB 158 H	Lincoln Green, later White	Converted to LWB, survives
35500011A	YVB 156 H	Tuscan Blue	Survives
35500012A	YVB 159 H	Bahama Gold	Survives
35500013A	YVB 169 H	Bahama Gold	Not known
35500014A	YVB 161 H	Bahama Gold	Survives
35500015A	YVB 162 H	Lincoln Green	Survives
35500016A	YVB 171 H	Lincoln Green	Survives
35500017A	YVB 163 H	Tuscan Blue	Survives
35500018A	YVB 164 H	Tuscan Blue	Survives
35500019A	YVB 170 H	Tuscan Blue	Not known
35500020A	YVB 172 H	Masai Red	Survives
35500021A	YVB 173 H	Davos White	Not known
35500022A	YVB 165 H	Davos White	Survives
35500023A	YVB 166 H	Masai Red	Survives
35500024A	YVB 167 H	Lincoln Green	Survives
35500025A	YVB 168 H	Lincoln Green, later Beige	Converted to LWB, survives
(LHD)			
35800001A	YVB 174 H	Sahara Dust	Converted to RHD and re-registered as 23 FWD; fate not known
35800002A	YVB 175 H	Sahara Dust	Survives
35800003A	Possibly YVB 176 H	Masai Red	Not known

The Launch Press Fleet

A batch of 20 right-hand drive vehicles was built and prepared for the press launch in June 1970. Some continued in use as press demonstrators and others were used by various departments at Solihull. Like the pre-production YVB-registered models, these vehicles have taken on an almost legendary status among Range Rover enthusiasts.

These vehicles had chassis numbers 35500026A to 35500045A and were registered consecutively from NXC 231 H to NXC 250 H.

Chassis	*Registration*	*Colour*	*Fate*
35500026A	NXC 231 H	Masai Red	Not known
35500027A	NXC 232 H	Masai Red	Survives
35500028A	NXC 233 H	Masai Red	Not known
35500029A	NXC 234 H	Masai Red	Survives
35500030A	NXC 235 H	Masai Red	Survives
35500031A	NXC 236 H	Davos White	Survives
35500032A	NXC 237 H	Davos White	Survives
35500033A	NXC 238 H	Davos White	Survives
35500034A	NXC 239 H	Davos White	Survives
35500035A	NXC 240 H	Davos White	Survives
35500036A	NXC 241 H	Tuscan Blue	Not known
35500037A	NXC 242 H	Tuscan Blue	Not known
35500038A	NXC 243 H	Tuscan Blue	Survives
35500039A	NXC 244 H	Tuscan Blue	Survives
35500040A	NXC 245 H	Tuscan Blue	Survives
35500041A	NXC 246 H	Lincoln Green	Survives
35500042A	NXC 247 H	Lincoln Green	Not known
35500043A	NXC 248 H	Lincoln Green	Not known
35500044A	NXC 249 H	Lincoln Green	Not known
35500045A	NXC 250 H	Lincoln Green	Survives

2 The First Nine Years (1970–1979)

The Range Rover was launched in June 1970 to wild acclaim. There was simply no other vehicle on the market which combined its levels of comfort and performance with its all-terrain ability and towing capacity. Its versatility was coupled with a great deal of character, and buyers flocked to buy the new vehicle.

By the end of 1970, Solihull knew that it had a runaway success on its hands. Demand had already outstripped supply, and before long nearly-new Range Rovers were changing hands at prices well above those demanded in the showrooms. Struggling to keep up, Rover made the vehicle available in European markets in the early part of 1971, but ironically never did launch it in the USA – the very market which had inspired its development. The main reason, undoubtedly, was that Rover could not afford to maintain their dealer network in such a huge country and actually pulled out of the US market in autumn 1971, at just about the time the Range Rover might have been launched over there.

This photograph shows all the characteristics of the very early Range Rover. Note in particular the round, screw-off filler cap, the rear window pillars painted in the body colour, and the bright metal framing on the rear side windows. Those round rear-view mirrors mounted on the bonnet would be standard equipment for several years.

These were bad times for Solihull, and the first few years of Range Rover production coincided with one of the most difficult and confused periods its makers had ever known. Absorbed into the British Leyland combine during 1968, Rover had maintained a degree of independence which was about to come to an abrupt end. During 1971, the Rover and Triumph engineering departments were merged. This was the prelude to a closer union between the two formerly separate companies in 1972, when Rover-Triumph (with a reluctant Jaguar) became the Specialist Division of British Leyland; at the same time, Austin, Morris and MG became the Volume Cars Division.

In themselves, these changes made little difference to the Land Rover side of the old Rover Company which was responsible for the Range Rover. However, the restructuring of British Leyland did destroy the financial independence of the individual marques, which were now made subject to a centralized financial control. This meant that profit-making divisions of British Leyland subsidized loss-making divisions, which might have been good business practice if only there had been enough profits to go round. In practice, the Volume Cars Division became so heavily subsidized by the Land Rover side of Rover-Triumph that there was no money left to be ploughed back into Land Rover products, which of course included the Range Rover.

Inevitably the crisis worsened. By the end of 1974, British Leyland was obliged to ask the Government for help. Nationalization followed, and the Government commissioned a report from industrialist Sir Don Ryder into British Leyland's future. Ryder's report was ready by the summer of 1975, and it contained a number of recommendations. Many of these were fairly quickly dismissed or discredited, but one which did survive proposed a major investment to increase production volumes of Land Rovers and Range Rovers. Unfortunately, it took three years before this recommendation could be implemented.

The result of all this was that no significant changes were made to the Range Rover for far too long. The development engineers were certainly working on improvements in the 1970s, but the money to put them into production was simply not forthcoming. There was a second problem, too: the Range Rover production lines had been squeezed into a corner of the South Works at Solihull and were already making as many vehicles as they could. Without further investment to establish a new production line, Land Rover could not even make as many vehicles as were being ordered. So for most of the 1970s, Range Rovers were in short supply and the original concept was not developed. If the Range Rover had not been so far ahead of its time when it was originally announced, it could not have survived for so long in such a difficult situation.

There was little to be done about the supply-and-demand problem, but something was done about Range Rover development. It was a well-established practice for Solihull to encourage aftermarket conversions of its Land Rovers to meet special needs, and exactly the same happened with the Range Rover. Solihull encouraged stretched-chassis, six-wheel, and special-body conversions for utility purposes from the beginning; but what they had not expected was a demand for more power and more convenience features. This came largely from the Middle East, where the Range Rover had found ready acceptance as upmarket desert transport, and where wealthy customers were prepared to spend thousands of pounds on having their vehicles custom-built. To meet demand, aftermarket converters sprung up to satisfy their every whim; in the process they carried out much of the development work which Solihull would have loved to do itself.

By the end of this period, the Range Rover had reached an impasse. Demand still exceeded supply, and there was no simple way of increasing production to keep up with demand. Similarly, it was obvious from the success of the aftermarket converters that there existed a demand for Range Rovers with automatic transmission, with four doors, with improved performance and with improved equipment. Yet Solihull simply did not have the facilities to develop and build such variants. It was already clear that Mercedes-Benz in Germany and several manufacturers in Japan were planning to plug the gaps which the Range Rover had left in the market with new products of their own. Something had to change if the Range Rover was to survive; fortunately, something did.

THE MARKETING

Solihull's Sales Division had trouble getting to grips with the Range Rover at first. For a start, they were bitterly disappointed that they had only a two-door model to sell instead of the four-door they were convinced the market would want. Secondly, they were selling an entirely new vehicle into an entirely new market, and were therefore obliged to invent a new marketing strategy.

Initial promotional material for the Range Rover capitalized on its versatility, which was perhaps its most readily identifiable feature. A cynic might also argue that by promoting it as all things to all men, the sales and marketing people could allow the vehicle to find its own market and then refine their promotional efforts to suit ... However, the sales and marketing people did pin their colours to the mast in one important respect right from the beginning. This was in promoting the Range Rover as an extension of the Rover car range rather than as an upmarket Land Rover. That, at least, was a shrewd move because it allowed the snob appeal and reputation for refinement of the Rover saloons to rub off onto the new vehicle.

Even so, sales material issued between 1970 and the end of 1972 made clear that the Range Rover was very much a working vehicle. In these early years, there was a suspicion among die-hard Land Rover customers that this new machine could not be as tough and rugged as the established 4x4 from Solihull, and the Rover marketing people tried very hard to alter that perception. At the end of 1971, they lent a pair of Range Rovers to the British Trans-Americas Expedition to make the first-ever north-to-south crossing of the American continent, and on their return made publicity capital out of the ruggedness which had enabled the vehicles to fight their way through the dense jungle of the Darien Gap in Central America. This was all grist to the mill, but the fact was that the Range Rover was already establishing its own image, independently of any manipulation by Solihull.

Wisely, Solihull listened to the feedback from the market-place, and in January 1973 the 'working vehicle' image was toned right down. The new sales literature issued that month spelled out that the Range Rover was now primarily a leisure vehicle:

> Leisure and pleasure mean different things to different people. For most of us the aim is to get away from the pressures of commercial and professional life, to do our own thing in places of our own choosing.

Versatility was a theme of early Range Rover advertising. These pictures illustrate well the estate-car configuration of the vehicle, with a folding rear seat which enabled extra loads to be carried.

> We may simply wish to exchange our own crowded environment for another more alluring one, or look for quieter pursuits where the air is fresher, where time matters less and the rumble of traffic is distant and unheard. You pay your money and you take your choice.
>
> With a Range Rover that choice becomes enormous .... This is ... the most versatile motor car in the world.

For the rest of the 1970s, the Range Rover continued to symbolize a lifestyle where wealth permitted an indulgence in outdoor leisure pursuits (even though relatively few owners actually lived up to that image). Sales were by this time so strong that there was no need to spend money on major advertising campaigns – and so for most of the 1970s the Range Rover was hardly advertised at all. Instead, it sold on its own reputation.

THE CUSTOMERS

'Who buys the Range Rover?' asked the first sales catalogue issued in 1970. 'Business and professional people with a leaning towards the great outdoors, who want a purpose-built vehicle instead of an adapted one, the real thing instead of a compromise.'

That, of course, was wishful thinking, because it had been written before the first Range Rovers went on sale. In practice, the Range Rover proved ideal transport for the wealthy country-dweller who needed a single vehicle for use on his estate and in town, and ideal transport for professionals whose work took them off the beaten track – architects, surveyors and civil engineers. However, the unifying factor among all these groups was money: the Range Rover cost as much as Rover's most expensive luxury cars in 1970, and as the decade wore on, its price actually increased.

Overseas, the picture was much the same. In countries like France and Italy, where large engines were heavily taxed, the Range Rover sold in small numbers to a very wealthy clientele. And in the Middle East, where the vehicle's combination of performance and off-road driving comfort gave it a special appeal for use in the desert, it once again became a favourite of the wealthy. Wherever it was sold, its size and high driving position made it the favoured transport of those who wanted to be noticed.

One way or another, the Range Rover certainly never lacked for custom during the first nine years of its existence. Sales gradually built up over the first five years, and by the mid-1970s it was consistently selling around 11,000 examples a year worldwide. If only production capacity had been greater, it would undoubtedly have sold in even greater quantities – and yet those low production volumes paradoxically also helped the Range Rover by minimizing the quantities available and thus giving it a certain exclusivity. Production for the UK market, for example, was limited to 2,500–3,000 examples a year for most of the 1970s.

THE PRODUCT, 1970–1979

There was something very special about a Range Rover. Whether it was the commanding driving position, the interior spaciousness, the feeling of superiority and invulnerability, or the combination of all three, there was no denying that this was a remarkable vehicle. And yet, by later standards, the first Range Rovers were quite frankly crude. The PVC seats, floor coverings and headlining may have been designed to be practical (in theory, the

Appearances can be deceptive! Although the original interior looked very neat when new, it did not wear well. The PVC upholstery quickly became hard and began to split, the floor covering of similar material tended to split around the gear-lever on the transmission tunnel and the door and bodyside trim panels came adrift. There was no outboard release lever for the front seat backs on these first models which also made life harder than it needed to be.

vehicle could be hosed out after a hard day's work), but they also looked and felt too utilitarian for a vehicle of the Range Rover's price and pretensions. Equipment was spartan, too, with minimal instrumentation and cheap-looking blanking plugs to cover the holes for additional instruments in the dashboard. The front seat releases were awkwardly placed, the gearchange was vague and agricultural, and both transmission and tyres created high noise levels.

However, it was hard in the beginning to know whether criticisms like these were valid or not. The Range Rover was a new type of vehicle and the customers could not compare it, favourably or unfavourably, with anything else on the market. So it was that Solihull did nothing at first to correct these shortcomings. In fact, the first changes were made as a way of saving costs. Jack Baines, who was then working in the Sales Division, remembers that an urgent meeting was held after the vehicle had been in production for a few months to see whether it was possible to cut the production costs. One change which resulted was that the black metal letters on bonnet and tailgate were changed for black plastic items. The accountants were no doubt also delighted to learn that Engineering proposed to delete the limited-slip feature from the centre differential because it chattered in use and offered almost no additional advantage.

For the moment, however, Solihull was caught up in a headlong rush to get the vehicle into all its intended markets and to introduce accessories and options. The first left-hand drive Range Rovers came off the lines at the beginning of 1971, and with them came a lower compression ratio for

Range Rover 1970–1979

Layout and chassis
Two-door estate with horizontally split tailgate
Box-section ladder-frame chassis

Engine

Type	Light-alloy 90-degree petrol V8
Capacity	3,528cc
Bore and stroke	88.9mm x 71.1mm
Fuel system	Two Zenith-Stromberg carburettors
Compression ratio	8.5:1 (1970), 8.25:1 (1971–7), 8.13:1 (1977–9)
Max. power	135bhp @ 4,750rpm (8.5:1 compression) 130bhp @ 5,000rpm (8.25:1 compression) 132bhp @ 5,000rpm (8.13:1 compression)
Max. torque	185lb/ft @ 2,500rpm (8.5:1 and 8.25:1 compression) 186lb/ft @ 2,500rpm (8.13:1 compression)

Transmission

Primary gearbox	4-speed LT95
Ratios	4.069:1, 2.448:1, 1.505:1, 1.00:1, reverse 3.664:1
Transfer gearbox	2-speed, integral with primary gearbox
Ratios	High 1.174:1 (1970–6), 1.116:1 (1976–9) Low 3.32:1
Centre differential	Lockable (with limited-slip feature on 1970 vehicles)
Final drive	3.54:1

Suspension and steering

Front	Live axle with coil springs and hydraulic telescopic dampers, located by radius arms and Panhard rod
Rear	Live axle with coil springs and hydraulic telescopic dampers, located by radius arms, support rods and central wishbone assembly incorporating a Boge Hydromat self-energizing ride-levelling strut
Steering	Recirculating-ball, worm and nut Power assistance optional from 1973
Ratio	Initially 18.2:1, later 20.55:1; 17.5:1 with power assistance
Tyres	205 x 16
Wheels	Pressed steel 16in Rostyle type

Brakes

Type	Disc brakes with dual hydraulic line and servo assistance; four-piston callipers on front wheels and two-piston type on rears; internal expanding drum-type parking brake operating on transfer box output shaft
Size	Front 11.75in diameter; rear 11.42in

Vehicle dimensions (in/mm)

Wheelbase	100/2,540
Overall length	176/4,470
Overall width	70/1,780
Overall height	70/1,780
Track	58.5/1,480
Kerb weight	3,800lb/1,723kg

Performance

Max speed	95mph/153km/h
0–60mph	14 sec

all engines. It seems possible that this lower compression was the one intended for the Range Rover from the beginning, but that supplies of the correct pistons had been delayed and so the first examples used pistons from the South African-specification Rover 3500 saloon. However, Geof Miller heard a different version of the story, which was that the engine people simply corrected a mathematial error in the calculation of the figure and that the actual compression ratio did not change at all!

The first options (which included fundamentals like a heated rear window and radio) did not go on sale until the spring, and it was at about this time that Solihull introduced a number of utility conversions, primarily for the emergency services (*see* Chapter 6). It was a very full schedule, and so the new model-year that autumn saw only essential corrections to the original specification. Further options were added in the early months of 1972 – as soon as they could be got into production – and even that autumn's new models for the

Solihull was planning interior improvements while the first vehicles were on sale. These experimental seats date from early 1972 and have prototype head restraints and fabric inserts. Their black colour never was approved for production and nor was the centre seat – actually taken from a Land Rover.

1973 model-year brought only minor changes. Not until the latter part of 1972 was Solihull able to start work on something really new.

The next round of changes came in January 1973, and they reflected the use to which the Range Rover was actually being put by its customers. Power-assisted steering was introduced, both to aid manoeuvring in car parks and to make women drivers feel more at home with a vehicle which had been essentially aimed at men. Customers who carried dogs in the back had complained that the ribbed flooring was uncomfortable for the animals, so a non-slip rubber mat was added; and in practice nobody ever hosed out the interior, so carpet was added to the transmission tunnel and a brushed nylon headlining was fitted, both to give a less spartan feel. The new headlining also had better sound-insulation properties than the original type.

Interior improvements figured large over the next couple of years. Brushed nylon upholstery replaced the original PVC, optionally at first and later as standard, and the transmission tunnel carpet was first extended to replace the rubber gear-lever gaiter and then upgraded. Inertia-reel safety belts also became optional, and represented a great improvement over the cumbersome static type originally fitted. Exterior cosmetics came in for attention in October 1973 as well, when black vinyl covers were added to the

Those enormous gaps between elements of the dashboard were quite typical of the times! This is a January 1973 model, with carpet on the transmission tunnel and auxiliary instruments filling all four spaces on the dashboard.

Spot the differences! This 1973-model Range Rover has a hinged fuel filler cap and the rear window frames are now black.

In January 1973 came a much-needed rear wash-wipe and black vinyl covers for the rear quarter-panels. That crude-looking number plate panel was designed to hinge down so that it would remain visible from behind if the vehicle were driven with the tailgate open.

rearmost window pillars. Vinyl roofs and pillars were then popular on cars of all kinds, and in the Range Rover's case they had the additional benefit of making the glass area look larger and thus making the vehicle look longer. However, the real reason for their introduction was quite different: the metal quarter-panels were very weak components and needed a lot of expensive hand working to make them fit in production, so the cheapest solution was to cover them up!

It was in 1973 that the first oil crisis of the 1970s erupted. The short-lived Arab-Israeli War of that autumn was followed by the Arab nations' decision to quadruple the price of crude oil and to embargo supplies to those nations they claimed had supported Israel during the conflict. Diplomatic negotiations eased the supply situation early in 1974, but crude oil remained expensive and the price of petrol at the pumps more than doubled. This enormous increase seriously affected the sales of all big-engined cars for a time, but the Range Rover continued to sell well, and sales for the 1974 season (that is, from September 1973 to August 1974) showed a 32 per cent increase worldwide over the previous season! No doubt the long

More differences – but not many. The Range Rover's evolution was painfully slow during the 1970s and this 1975 model differs from earlier types mainly by its scuttle side badges, which were now printed instead of engraved. There is also a larger space between the words and the edge of the badge plate.

waiting-lists for the vehicle helped cushion the effect of the oil crisis, but the Range Rover also survived so well because many of its buyers were wealthy enough not to be unduly concerned by increases in the price of fuel.

Solihull did respond to a social climate in which heavy fuel consumption had become something of a moral issue as well as a practical one, but that response came very late. Not until October 1976 (more than two and a half years after fuel economy had become an issue) did the company raise the High Ratio in the Range Rover's transfer box by 5 per cent to give taller gearing for road use, while adding a more efficient twin-pipe exhaust system. Two years after that came the option of a Fairey overdrive (which had been available for Land Rovers since 1974), although this was not a likeable feature because it added one more ugly lever to the Range Rover's transmission tunnel and contributed an extra cacophony of clunks, bangs and gear whine to its progress. Even with all these changes, the Range Rover was not noticeably more economical in everyday use. Only in fast motorway cruising was there a real difference, and in such conditions, *Autocar* magazine recorded 16–17mpg (17–18l/100km) with an overdrive Range Rover in July 1978, as compared to 14–15mpg (19-20l/100km) for the standard vehicle.

Other considerations affected the way the Range Rover developed in the middle and later 1970s, too. Industry trends led to the arrival of door mirrors in place of the bonnet-mounted type for the 1978 models, and to matt black windscreen wiper arms in place of silver ones a year later. The 1978 models introduced in October 1977 also had a lower engine compression ratio of 8.13:1, although revised camshaft profiles meant that the engine was actually marginally more powerful than the older 8.25:1 compression type. In this case, the change had been made because of new exhaust emissions regulations in the Australian market, where Range Rovers were selling well, and the low-compression pistons were standardized for all markets in order to simplify engine production.

Look very hard and you might just spot the differences on the 1978-model Range Rover. The rear-view mirrors are now on the doors and the cloth upholstery is just visible through the windows. The foglamps under the bumper were not standard equipment, but came from the approved options list.

Special Export Specifications

Market requirements and local regulations ensured that some export versions of the Range Rover differed in detail from the 'standard' or Home Market version.

Thus, for example, Range Rovers for Sweden and Australia always had fixed front head restraints, while Germany, Norway and Sweden had a 'detoxed' (emissions-controlled) engine with the Pulsair air injection which did not become standard elsewhere until 1981. Australia also took a similar engine after 1977.

German Range Rovers always had an adjustable backrest for the driver's seat (which was never available anywhere else), and safety regulations in that country also demanded two further special features towards the end of the 1970s. These were fairings ahead of the exposed bonnet hinges and plastic bumper end caps both front and rear. The rear end caps incorporated a red reflector, and the reflectors in the rear lamp units were covered over with small metal plates bearing a 'V8' badge.

Some territories demanded side repeater flashers towards the end of the 1970s, and these were of the type fitted as standard after autumn 1979. In addition, an in-dash air conditioning system became available some time around 1978 for some export territories, although it could not be had on Home Market models until autumn 1979.

Key Production Changes, 1970–1979

NB Paint and trim colour option changes are listed separately.

June 1970	Range Rover announced
December 1970	Limited-slip feature deleted from centre differential
February 1971	Engine compression lowered from 8.5:1 to 8.25:1
May 1971	First options: Towbar Trailer electrics Fog lamps Long-range driving lamps Radio Sundym tinted glass Laminated windscreen Heated rear window
October 1971	New options: Uprated (18 ACR) alternator Split charging system with second battery Roof rack
January 1972	New option: Cigarette lighter
April 1972	Ammeter, oil pressure gauge and oil temperature gauge added to blanked-off holes in facia centre Changed option: Wider choice of radios
October 1972	Steering ratio lowered from 18.2:1 to 20.55:1 Gearchange improved Hinged fuel filler cap replaced screw-on type Body structure strengthened Changed Option: Radio choice reduced to one New options: Front mudflaps

	Rear safety belts Safety snap-stem interior mirror Fuel tank guard plate Centre PTO
January 1973	Rear wash/wipe added Outboard release handles for front seats Carpet replaced PVC matting on transmission tunnel Brushed nylon headlining replaced vinyl type Single-hand-release safety belt catches Non-slip mat for rear loadspace Uprated (18 ACR) alternator now standard New option: Power-assisted steering
October 1973	Black vinyl coverings for rear quarter-pillars Printed scuttle side badges replaced engraved type Differential lock warning light moved from transmission tunnel control to dashboard Thermostatically-controlled intake for engine air cleaner Improved mountings for tailgate gas struts Stronger channels for sliding rear windows Electric fuel pump now standard (vertical mounting replaced horizontal type) New options: Brushed nylon seat facings Inertia-reel safety belts
October 1974	Carpet gaiter for gear lever replaced rubber type
October 1975	Improved carpeting on transmission tunnel Two roof-mounted courtesy lights instead of one Modified carburettors Changed option: Split charging system now had uprated (20 ACR) alternator New option pack consisting of: Brushed nylon seat facings Front seat head restraints Inertia-reel safety belts (on front seats only) Sundym tinted glass Power-assisted steering
October 1976	Transfer box High Ratio raised from 1.174:1 to 1.116:1 Twin-pipe exhaust replaced single-pipe type Changed option: Split charging system now had further uprated (25 ACR) alternator
October 1977	Engine compression ratio lowered from 8.25:1 to 8.13:1 Door-mounted rear view mirrors replaced bonnet-mounted type
October 1978	Black wiper arms, front and rear, replaced aluminium-finish items New option: Fairey overdrive

Paint and Trim Options, 1970–1979

June 1970 to September 1973
There were six standard paint options. All trim was in Palomino PVC.
The paint options were:

Bahama Gold	Masai Red
Davos White	Sahara Dust
Lincoln Green	Tuscan Blue

October 1973 to September 1974
There were six standard paint options. Standard trim was Palomino PVC but Bronze brushed nylon trim was optional. The standard paint options were:

Bahama Gold	Masai Red
Davos White	Sahara Dust
Lincoln Green	Tuscan Blue

October 1974 to August 1979
There were six standard paint options. Standard trim was Palomino PVC but Bronze brushed nylon trim was optional. The standard paint options were:

Arctic White	Masai Red
Bahama Gold	Sahara Dust
Lincoln Green	Tuscan Blue

Chassis Numbers and Production Figures, 1969–1979

These figures show the build date, by calendar year, of Range Rovers built between 1969 and 1979. They were calculated by Anders Clausager of the British Motor Industry Heritage Trust from the factory Despatch Records which the Trust holds. They are used by kind permission.

1969–1975
Range Rover chassis were numbered in three separate series, identified by three-figure prefixes: 355 for RHD Home Market models, 356 for RHD Export versions, and 358 for LHD models (357 was reserved for RHD CKD and 359 for LHD CKD, but no records are available).These identifiers were followed by five-digit serial numbers and by a letter suffix which indicated major production changes of importance when servicing the vehicle. A typical chassis number would therefore be 35501238A, which belonged to a 1971 Home Market model.

	355	356	358
1969	00001–00003	None built	None built
1970	00004–00312	00001–00005	00001–00006
1971	00313–03157	00006–00068	00007–00745
1972	03158–05718	00069–00820	00746–03227
1973	05719–08659	00821–01857	03228–05837
1974	08660–10572	01858–03156	05838–09850
1975	10573–11062	03157–03292	09851–10556

1975–1979

From approximately February 1975, Range Rover chassis were numbered in a single series. However, the five-digit serial numbers were prefixed by the same three-figure identifiers as before and were again followed by letter suffixes. It is not clear why the number 12024 was chosen for the start of the new sequence.

1975	12024–21662
1976	21663–31094
1977	31095–40479
1978	40480–55741
1979	55742–61821

(The 1979 figures run to October 1979 approximately. In theory, the 1980-model vehicles were all numbered under the new system introduced at the time of their launch, but there may have been some vehicles built to 1980 specification with old-style chassis numbers.)

3 Beginning the Turnaround (1979–1984)

Over the next five years, Land Rover finally tackled the Range Rover's most glaring shortcomings, elevating it from an excellent but underdeveloped vehicle and unlocking hidden potential which even its original designers may never have suspected was there.

Yet it was only just in time. Recognizing that those shortcomings had left a gap in the market, Mercedes-Benz had developed the G-Wagen as a Range Rover competitor, and they launched it at the Frankfurt Show in the autumn of 1979. In Japan, meanwhile, Isuzu and Mitsubishi had also spotted a gap, this time for a vehicle which offered most of what the Range Rover had but cost rather less, and it was in 1981 and 1982 respectively that the Trooper and Shogun were launched on world markets. Under their impact and that of later arrivals, the 1980s would become the decade of the four-wheel drive estate, and it was only through constant development that Land Rover were able to keep the Range Rover on top of the heap during that period.

THE CHANGING FACE OF THE COMPANY

The catalyst for the major changes which were made to the Range Rover between 1979 and 1984 was a vast organizational change at Solihull. This change established Land Rover Ltd as a separate operating company within British Leyland, and provided it with an investment budget which allowed the wrongs of the earlier 1970s to be righted over a period of five years. It was a period of intense activity both for the Range Rover's manufacturers and for the development of the vehicle itself; and it provided the Range Rover with a secure basis for more than a decade of further production.

The story of these organizational changes goes back to 1975, when the report on British Leyland's troubles which Sir Don Ryder produced for the Government recommended massive investment to increase Land Rover and Range Rover production. Many elements of the Ryder Report were quickly discredited, but this recommendation remained intact, and during the middle 1970s a total of £400 million was earmarked for investment in new manufacturing facilities.

Meanwhile, the British Leyland combine was going through its darkest period. It was not until November 1977 that Michael (later Sir Michael) Edwardes was appointed as its Chairman, with a three-year contract which in effect gave him the task of making or breaking the company. One of his first actions was to break down the amorphous Leyland Cars into a series of smaller divisions with greater autonomy, and he also recognized the value of retaining the marque names which Leyland Cars had been in the process of suppressing. So it was that the Large Cars division became Jaguar Rover Triumph, and within that Edwardes established Land Rover Ltd as a company on its own.

Land Rover Ltd came into existence in July 1978, and it was given its own budget and its own Managing Director. That Managing Director was Mike Hodgkinson, and one of his first tasks was to review the £400 million budget which had been earmarked for the Solihull 4x4s under the Ryder Plan. Hodgkinson decided to take a more cautious approach. He believed that a simple increase in production volume was not the answer; once current demand had been satisfied, there was every chance that orders would tail off unless new products had been developed to bring in renewed orders. He therefore recommended a lower level of investment in Land Rover, which would be split between new production facilities and the development of both existing and new products.

Hodgkinson's plan called for just £280 million to be invested in Land Rover over a period of five years. The first stage would cost £30 million and had already been partially implemented when the plan was made public in August 1978. As restricted supplies of V8 engines were one reason why Range Rover production had not been increased earlier, 'Stage I' of the plan paid for a new V8 engine assembly line to go into the Acocks Green engine plant. This would not only allow more Range Rovers to be built, but would also permit the introduction of the V8 engine in the Land Rover. (This stage of the investment plan also gave its name to the the V8 Land Rover introduced in 1979, which became known as the Stage I model.)

Stage I also paid for retooling at the Tyseley gearbox plant, and as far as the Range Rover was concerned it had two very important effects. The first was that production volumes could increase by 50 per cent to 450 vehicles a week. The second was that the vehicle could be freshened up with a long-overdue series of revisions. Land Rover had these ready for public announcement in September 1979.

It was Stage II which consumed the lion's share of the new investment in Land Rover, however. Its £250 million went towards the introduction of more new products – notably the coil-sprung Land Rover One Ten in 1983 – but were also used to finance a whole series of changes to the Range Rover. The old North Works at Solihull, which had been empty since Rover P6 saloon car production had ceased in 1976, was completely refurbished and extended at a cost of £85 million. The work began in 1979 and was finished early in 1981, when the Range Rover assembly lines moved into the new purpose-built plant.

The value of the new assembly plant to Range Rovers was enormous. It gave the Solihull flagship a self-contained plant and paint shop which covered some 565,000sq. ft, and provided the capacity to build as many as 600 vehicles a week on a two-shift system if required. The lines were also equipped with more modern machinery than before and their layout permitted a production flexibility which had simply not been possible with the old Range Rover line. The first outward sign of this was the four-door Range Rover, which was introduced just a few months after the new North Works had opened. Both four-door and two-door models were built on the same lines, which allowed Solihull to vary the proportions of one to the other to suit fluctuations in market demand.

Without that flexibility, Land Rover would have been unable to introduce the four-door body, the automatic transmission, and the wide variety of optional specifications with which Range Rover appealed over the next three years to an even broader market than before. Most of the credit for the direction which Range Rover development took in the early 1980s must therefore go to Mike Hodgkinson, whose foresight in 1978 laid the foundations for Land Rover Ltd. Although Hodgkinson left Land Rover at the end of 1982, to be replaced by Tony Gilroy, his five-year plan still had more than a year to run. It was only after that, as the next chapter shows, that Gilroy began to exert his own very powerful influence over the Range Rover's destiny.

The establishment of Land Rover Ltd was of course only one of the many steps which helped British Leyland as a whole to recover from the disastrous 1970s. By the early 1980s, things were once again beginning to look good for the company, but the stigma of the name lingered on. So in mid-1982, as part of another reorganization, British Leyland was renamed the Austin Rover Group. This resulted in a change to the name which appeared on the manufacturer's plate in every Range Rover, but neither the vehicle itself nor Land Rover Ltd was otherwise materially affected.

THE MARKETING

There were several changes to Range Rover marketing as the 1980 models came on stream. For a start, promotional material acknowledged the vehicle's links with Land Rovers – as indeed it had to now that the vehicle was being built by Land Rover Ltd. The Range Rover was no longer marketed alongside Rover saloons, which was perhaps a good thing in view of the poor reputation these had acquired by the end of the 1970s.

Style would be of increasing importance in Range Rover marketing over the next few years. Limited editions and special editions were used to add an aura of exclusivity to Range Rover ownership (although they also served other product-development purposes) and the vehicle was gradually moved further up-market into the lower reaches of the luxury-car sector. Much of this was dictated by the arrival in world markets of cheaper four-wheel drive estates from Japanese manufacturers; Land Rover capitalized on the Range Rover's snob appeal, which the Japanese vehicles simply did not have.

The Range Rover also had one more powerful advantage over its Japanese imitators. Whereas they were mostly unknown quantities in the four-wheel drive market, the Range Rover could point to its links with Land Rovers and draw some credibility from them. It was a fine balancing act for the marketing men, but they succeeded only too well. For the 1983 model-year, a highly effective new slogan neatly evoked the Land Rover's association with safari and expedition work and linked this with the Range Rover's more up-market qualities. That slogan was 'Luxury need not stop where adventure begins', and it continued in use for three more years.

THE CUSTOMERS

It went almost without saying that the typical buyer of a new Range Rover in this period was fairly wealthy: the vehicle always was expensive, and became more so as time went by. In Britain, the vehicle had particular appeal to country-dwellers, although as the specification became more luxurious, so increasing numbers were bought by those living in towns.

Land Rover's own market research also showed certain other trends. Those interested in outdoor pursuits, such as hunting or ballooning, were buying Range Rovers as support vehicles for their hobbies. Those whose work involved carrying goods also liked the vehicle. And above all, those who needed a powerful and stable tow-car bought Range Rovers. The customer profile was very much the same as the one Graham Bannock's research had identified back in the mid-1960s; the main difference was the increasing numbers of people who were buying it for the image it presented rather than for the reality of needing it. In many countries, it was becoming commonplace for a Range Rover owner never to take the vehicle off the road, and to be no more than dimly aware of its formidable off-road ability.

THE PRODUCT, 1979–1981

Although the Managing Director of a company deserves the credit for the overall strategy which makes a product successful, the credit for the design and development of that product lies elsewhere. During this period of the Range Rover's production life, Tom Barton was Chief Engineer at Land Rover, the man who co-ordinated teams of engineers to devise the product the company wanted. Styling, meanwhile, was the responsibility of British Leyland's Styling Director David Bache, who had come a long way since he had been the Chief Styling Engineer at Rover who had tidied up the Range Rover prototypes for production. In practice, most Range Rover styling changes of this period were delegated to a team under Tony Poole, who remained the most senior Land Rover stylist after Bache left the company in 1981.

Land Rover's own market research must have given a strong steer to those involved with the Range Rover when they came to decide how the vehicle should be developed. However, it was also true that the aftermarket converters had already developed the Range Rover to suit some types of demand, and no doubt Land Rover saw little point in trying to re-invent the wheel. They therefore looked at what the converters had done and, where appropriate, they borrowed ideas and developed them further to ensure that they met Land Rover Ltd's own performance criteria. For that reason, several of the 'new' features which appeared on Range Rovers in the next few years seemed naggingly familiar to those who had followed the Range Rover's career at all closely. And for that reason, Land Rover also entered into limited contractual agreements with a small number of the conversion specialists.

The minor changes made to the Range Rover in October 1979 were irrelevant in this context; they were in any case a low-budget operation and their primary purpose was to draw attention to the improved availability of Range Rovers which resulted from increased production. Nevertheless, they did make the Range Rover a better and more desirable vehicle, as did the additional cosmetic changes introduced rather less than a year later; but they also made the vehicle more expensive and thus less attractive to fleet users and some conversion specialists. So in order to retain their fleet market, Land Rover introduced a cheaper, less well-equipped Range Rover variant called the Fleet Line during 1980.

The October 1979 changes introduced very little that was actually new; in most cases, it was simply that formerly optional items had been made standard. However, the overall result was inevitably to upgrade the vehicle. A few small distinguishing touches also helped to put across the message that the bad old days of the Range Rover's stagnation were over. Black

The Range Rover Fleet Line, 1980–1984

Total built	Not known
Special features	Manual steering Pleated Ambla upholstery Rubber floor mats instead of carpets

The Range Rover Fleet Line was introduced in 1980 when Land Rover found that many fleet users (notably UK Police Forces) and the converters who catered for the commercial and utility markets were unwilling to pay the extra which the latest upgraded specification demanded. It was in essence a 'stripped-out' version of the contemporary two-door model. The Fleet Line's popularity gradually declined as fleet users took to the four-door models, and it was effectively dead by 1984.

bumpers and repeater lamps on the front wings were the obvious signs; new paint finishes freshened up the nine-year-old shape; and tape badges front and rear made cleaning easier as well as reducing the rust hazard presented by the fixing holes for the earlier badges. Probably the most significant new feature was the wider availability of the integrated air conditioning system which a few overseas markets had seen a year or so earlier, and which replaced the roof-mounted type fitted to early vehicles. In the UK market, it was an option which remained rare, but it did make a valuable difference to Range Rover sales overseas.

The changes made for the 1981 models announced in June 1980 were even less far-reaching, focusing primarily on the interior. Nevertheless, a new brushed velour upholstery did help to make the Range Rover feel more luxurious, and thus represented a step in making the vehicle match

Instant recognition features of the 1980-model Range Rovers when they were new included black bumpers, new decal badging and the indicator repeater on the front wings (although this had been on earlier models for some overseas markets). The standard tyres also had a less aggressive tread pattern than those originally fitted.

New for the 1980 model was a four-spoke steering wheel; cloth upholstery and front head restraints also became standard after being optional for several years. However, the controls were still an ergonomic mess and the build quality of the dash-board had not improved since the earliest days.

customer expectations. But the really important changes which Solihull had in the pipeline did not start to appear until 1981. The opening of the new North Works assembly plant at the beginning of that year removed one of the biggest hindrances to the Range Rover's progress by providing flexibility on the production lines. The new models which were to result from that flexibility would quickly satisfy most of the outstanding customer dissatisfaction with the original vehicle.

Very much more was going on behind the scenes, however. This full-size styling buck was pictured in November 1980, when it was being used for trials of new front-end styling; proprietary styled wheels were also under consideration.

Meanwhile, Land Rover prepared the market for what was to come with a series of three special Range Rover variants which were sold alongside the regular production models. These three variants also helped to confirm that the company's customer research had been correct by testing the water for the planned full-production models. By developing each variant in conjunction with one of the aftermarket conversion specialists, and by making them available in only limited numbers, Land Rover minimized their own expenditure and contained the risk of failure. And at the same time, the restricted availability of these variants added to the aura of desirability which already surrounded the Range Rover.

The Range Rover Monteverdi tested the market for a four-door model. The body conversion was actually carried out by Fissora, an Italian company with links to Monteverdi.

The first of these special variants was introduced at the Geneva Motor Show in March 1980, and was a four-door conversion by Automobiles Monteverdi of Basle in Switzerland. Solihull had already granted Approval to the four-door conversion by FLM Panelcraft, but this was a rather ungainly vehicle and Land Rover were probably never really happy with it. On the other hand, Monteverdi's conversion was neat and slick. The official version of the story is that Monteverdi had developed the conversion and approached Land Rover Special Projects for approval, probably during 1979. Land Rover were then said to be so impressed that they agreed not only to grant Approval but also to sell the Monteverdi conversion through their own dealerships as an extension of their own range.

Whether Monteverdi knew that Land Rover were already developing their own four-door body is not clear. What does seem clear, however, is that the Range Rover Monteverdi never sold in anything like the volumes anticipated. The original plan was for Monteverdi to convert about 300 vehicles a year, although production was halted when Land Rover introduced their own four-door Range Rover just sixteen months after the Monteverdi had been announced, and it appears that no more than fifty examples had been built by that stage. Of

The Monteverdi incorporated a more luxurious interior than the standard type. Leather upholstery, special door trims and the neat cowling over the instrument panel are all visible in this picture. Note also the shut-line of the rear door, which was angled forwards and then ran along the leading edge of the wheelarch.

these, thirty-five were sold through Land Rover dealers in the UK and the other fifteen in Europe; the last few were not put on the road until 1982.

The Monteverdi vehicles were converted in Italy from fully-assembled two-door Range Rovers which were shipped from Solihull finished in white primer. In addition to the four-door body, they had several interior refinements (such as optional leather upholstery) and thus helped Land Rover to assess the market for more luxurious Range Rover variants. Monteverdi showed a turbocharged four-door model at the Geneva Show in 1981, but this may have been a one-off. The company also trimmed a few factory four-doors in 1981–1982, apparently for private customers, and in 1980–1981 produced a Range Rover design study for Land Rover. Nothing came of this exercise.

The second special variant was developed with Schuler (*see* Chapter 7), and appears to have become available shortly after the Monteverdi in 1980. The purpose of this one was to assess the market potential for a Range Rover with automatic transmission. Schuler had already developed an automatic-transmission conversion, which used an American Chrysler Torqueflite three-speed main gearbox and Schuler's own chain-driven transfer box. This was linked to a Ferguson Formula anti-skid braking system (developed by Schuler in conjunction with FF Developments), and the whole conversion was fitted to a batch of twenty-five vehicles for Land Rover. In other respects, these vehicles were standard.

Schuler no doubt hoped to interest Land Rover in their transfer box and anti-skid brakes, but it was not to be. Solihull had its

The Range Rover Monteverdi

Total built	50 (figure disputed)
Special features	Four-door body conversion by Automobiles Monteverdi of Basle, Switzerland
	Air conditioning
	Beige, Black or Tan cloth upholstery; cloth standard, pleated leather optional
	Carpeted rear load area
	Stowage box on transmission tunnel
	Fairing over instrument binnacle
	Special plate badges (at least two different types were used)
	Special paint finishes (Light Blue metallic, Light Green metallic, Anthracite metallic, Silver Grey metallic and White were listed; but in practice probably any colour could be had to special order)

The Monteverdi was used to test the market for the proposed factory-built four-door model and for more luxurious Range Rovers in general. It was sold through Land Rover dealers and was listed as the most expensive model of the range in 1980–1981. Production stopped when the factory-built four-door was introduced in 1981. As the Monteverdi was essentially a custom-built vehicle, it is probable that no two examples were exactly alike.

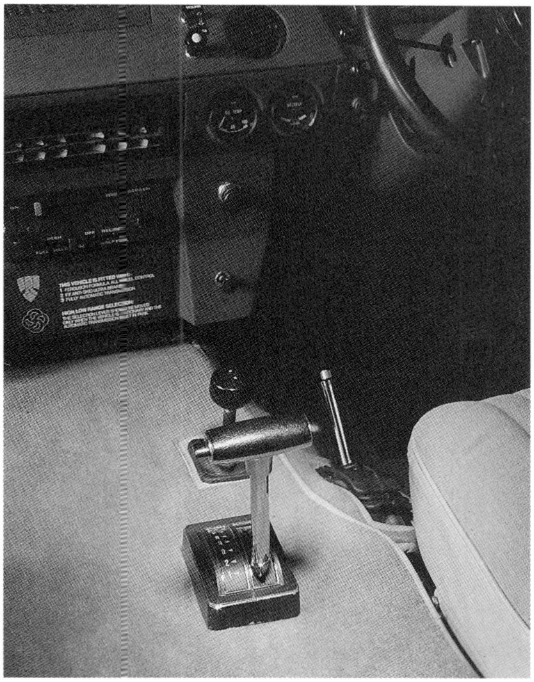

Inside the special-edition Schuler Automatic was a special plate with information about the transmission. The carpet on the transmission tunnel was finished much more neatly than in line-production vehicles.

own new transfer box under development by this stage, and was unwilling to go for the complication and expense of anti-skid brakes. By the summer of 1981, prototypes of Land Rover's own Range Rover Automatic – which used the same Chrysler gearbox – were running, and the Schuler link came to nothing.

The third special Range Rover variant was built in far larger numbers, although it was limited to 1,000 to give it a certain exclusivity. Towards the end of 1980, Land Rover worked with Wood and Pickett to design a specially-equipped luxury two-door model. The prototype (HAC414W) was then lent to Vogue magazine, who took it to Biarritz and used it as the backdrop in photographs of the 1981 Lancôme and Jaeger fashion collections.

When these pictures appeared in the magazine, they immediately aroused interest in the vehicle. Land Rover were ready with production examples, and the Range Rover In Vogue went on sale in February 1981. In fact, the production examples had several differences from the prototype. They had standard wheels instead of three-spoke alloys (although these would become a Range Rover option later in 1981), they had a different style of coachline along the body sides, and they may also have been a slightly different colour! The production metallic light blue paint and the broad two-tone grey coachline chosen for production, incidentally, were both also seen during 1981 on some versions of the then-new Austin Mini Metro.

Sales appear to have been fairly brisk, and it is doubtful whether many of these vehicles remained in the showrooms by the

The Schuler Automatic, 1980

Total built	25
Special features	Chrysler A727 Torqueflite three-speed automatic transmission Schuler-Voith chain-driven transfer box Ferguson Formula anti-skid brakes

The Schuler Automatic was intended to test the market for the proposed factory-built Range Rover Automatic. It was built for Land Rover by Schuler Presses of Sunningdale, Berkshire. The vehicles were sold through Land Rover dealers but were never advertised.

summer of that year. The In Vogue limited edition no doubt provided further valuable feedback on the market for a luxuriously-equipped Range Rover, and it also performed a valuable service in marketing the vehicle.

The In Vogue Two-Door Model, 1981

Total built	1,000
Special features	Light blue metallic (Vogue Blue) paint with broad twin coachlines in two-tone grey
	Black centre caps for wheels
	Air conditioning
	Polished wooden door cappings
	Stowage box between front seats
	Map pockets on front seat backs
	Fully carpeted load area
	Carpet material for spare wheel cover and tool kit curtain
	Picnic hamper in rear loadspace
	Stainless steel tailgate capping
	High-compression (9.35:1) engine
	Transfer box with taller (0.996:1) high ratio

This first limited-edition Range Rover was designed in conjunction with Wood and Pickett of London (*see* Chapter 8). All examples had right-hand drive and all were sold in the UK. The prototype (which had three-spoke alloy wheels) appeared in a *Vogue* magazine fashion feature, and the name In Vogue came from this promotional link. The model was used to test the market for more luxurious Range Rovers, and it also pioneered a number of features which would appear later that year on production vehicles: the high-compression engine and new transfer box gearing, the central stowage box and seatback map pockets, the carpeted load area, centre caps for the wheels, and metallic paint.

The In Vogue limited edition was created with a little help from customizing experts Wood and Pickett. Apart from the light metallic blue paint, the major distinguishing features were the twin coachline and wheel centre badges.

THE PRODUCT, 1981–1984

All three of these limited-edition variants, fascinating though they were, are nevertheless peripheral to the main thread of the Range Rover story. In July 1981 the company announced what was undoubtedly the most important development of the Range Rover since its introduction in 1970 – the option of a four-door bodyshell.

Even before the first Range Rovers had been built, the Rover Sales Department had argued that the market wanted a four-door model rather than a two-door. However, Rover was then an engineering-led company, and the engineers had got their way arguing that the two-door shell was simpler to design and build. They did build a four-door prototype in the early part of 1972 as a feasibility study, but shelved the project because money was tight and did not return to the charge until 1978. Gordon Bashford, among others, was involved in sketching up proposals for the four-door body, and it is clear that serious consideration was given to stretching the wheelbase to around 109in for the four-door model. However, cost eventually dictated that the company should stay with the standard wheelbase.

The factory-built four-door body was neater and simpler even than Monteverdi's conversion. Its main advantages were a straight shut-line for the trailing edge of the rear doors in place of the complicated curves of the Monteverdi, and a smaller quarter-light in the front door, which made the door look larger. In addition, the windscreen pillars were finished in black satin paint to make the glass area look larger. Morris Marina door handles, exactly like those used on the 1972 prototype (the Monteverdi's had come from the Austin Allegro) replaced the edge-pulls of the two-door models, and the only jarring note was perhaps the keyhole on the front doors, which looked rather lost half-way down the trailing edge of the panel.

When Solihull's own four-door model arrived in 1981, it sported a straight shut-line at the edge of the rear door and a smaller opening quarter-light.

Range Rover 1979–1984

Layout and chassis

Two-door estate with horizontally split tailgate (1979–84), or four-door estate with horizontally split tailgate (1981–4). Box-section ladder-frame chassis

Engine

Type	Light-alloy 90-degree petrol V8
Capacity	3,528cc
Bore and stroke	88.9mm x 71.1mm
Fuel system	Two Zenith-Stromberg carburettors
Compression ratio	8.13:1 (1979-1981) 9.35:1 (1981-1984)
Max. power	132bhp @ 5,000rpm (8.13:1 compression) 125bhp @ 4,000rpm (9.35:1 compression)
Max. torque	186lb/ft @ 2,500rpm (8.13:1 compression) 185lb/ft @ 2,500rpm (9.35:1 compression)

Transmission

Primary gearbox	4speed LT95 (1979–83) 5-speed LT77 (1983–4) Optional three-speed Chrysler A727 automatic, 1982–4
Ratios	4-speed: 4.069:1, 2.448:1, 1.505:1, 1.00:1, reverse 3.664:1 5-speed: 3.32:1, 2.13:1, 1.39:1, 1.00:1, 0.77:1, reverse 3.42:1 Automatic: 2.45:1, 1.45:1, 1.00:1, reverse 2.20:1
Transfer gearbox	2-speed, integral with primary gearbox on 4-speed models 2-speed LT230R on 5-speed and automatic models
Ratios	High 1.116:1 (1979–80), 1.003:1 (1980–1 and with automatic transmission, 1982–4), 0.996:1 (1981-3 with 4-speed primary box), 1.19:1 (with 5-speed primary box) Low 3.32:1
Centre differential	Lockable
Final drive	3.54:1

Suspension and steering

Front	Live axle with coil springs and hydraulic telescopic dampers, located by radius arms and Panhard rod
Rear	Live axle with coil springs and hydraulic telescopic dampers, located by radius arms, support rods and central wishbone assembly incorporating a Boge Hydromat self-energizing ride-levelling strut
Steering	Power-assisted recirculating-ball, worm and nut
Ratio	17.5:1
Tyres	205 x 16
Wheels	Pressed steel 16in Rostyle type; three-spoke alloy type on some models

Brakes

Type	Disc brakes with dual hydraulic line and servo assistance; four-piston callipers on front wheels and two-piston type on rears; internal expanding drum-type parking brake operating on transfer box output shaft
Size	Front 11.75in diameter; rear 11.42in

Vehicle dimensions (in/mm)

Wheelbase	100/2,540	
Overall length	176/4,470	
Overall width	70/1,780	
Overall height	70/1,780	
Track	58.5/1,480	
Kerb weight	Two-door models	3,872lb/1,756kg
	Four-door models	3,942lb/1,788kg

Performance

Max speed	95mph/153km/h
0–60mph	14 sec

The Land Rover team had also done an excellent job inside. Redesigned door trims left no metal exposed and added to the aura of refinement, while the rear seat had been moved back by 3in (75mm) to give extra leg-room. Between the front seats was the stowage box first seen on the In Vogue earlier that year, but Land Rover had not yet been able to develop a safety-belt mounting on the central pillar; so, for the time being, the front seats were identical to those in the two-door models, with integrally-mounted safety belts and no provision for backrest adjustment.

Despite such shortcomings, the four-door Range Rover went down extremely well with the buying public. It also won a Design Council Award in 1982 for 'excellence in design and in recognition of an outstanding product from British industry'. Land Rover had predicted that the four-door would eventually account for four out of every five Range Rovers sold; within a year that figure had very nearly been reached when 75 per cent of all new vehicles were four-doors; and within three years the four-door was taking 90 per cent of sales on the UK market.

With the arrival of the four-door came several minor refinements which affected the two-door model as well, but from now on two-door models would always lag behind the four-door in equipment levels. Only on the four-door did it become possible to order electric window lifts and metallic paint, but both two-doors and four-doors could be had with the smart three-spoke alloy wheels first seen on the In Vogue prototype in 1980. These, remembers Tony Poole, were known in the Styling Department as the Mustang wheels because their basic shape had been copied from the wheels of the 1979 Ford Mustang: obliged to come up with an alloy wheel design in a hurry when the Marketing Department insisted alloy wheels would help sales, the Land Rover stylists had done the best they could!

The Trophy Range Rover

During 1981, Land Rover became involved with the Camel Trophy adventure rally for the first time, and the company provided competing teams with two-door Range Rovers. At this stage, the Camel Trophy was open only to teams from West Germany.

The German Range Rover importers, Leyland GmbH, capitalized on the vehicle's involvement in this event by announcing a limited-edition Trophy Range Rover, shortly after the event was over. The vehicle's specification included a bull bar, additional lamps, compass, map-reading lamp, distance recorder and other expedition-oriented extras.

(The Camel Trophy is covered in greater detail in Chapter 11.)

Both two-door and four-door Range Rovers now had the high-compression engine and taller High Ratio transfer box gearing which had been previewed on the In Vogue model earlier in the year. This package of changes was an ingenious one which reduced noise and improved fuel economy without affecting acceleration or top speed.

The higher compression of 9.35:1 (shared with the contemporary Rover 3500 saloon engine) contributed to better fuel economy, while a new low-lift camshaft brought different valve timing and lift, and allowed the engine to reach maximum power at 4,000rpm instead of 5,000rpm. That reduced noise levels at high speed and further reduced fuel consumption. Maximum power was down, from 132bhp to 125bhp, but as the Range Rover's top speed was limited by its barn-door aerodynamics rather than power, the difference went unnoticed. Meanwhile, the new camshaft also raised maximum torque and gave more low-down torque, and this in turn allowed the transfer gearing to be raised for even better fuel economy. Different carburettors were fitted to suit the revised engine characteristics, and a new sliding-contact distributor enabled the engine to remain in tune over higher mileages than before.

When air conditioning was fitted, there was a different lower facia and the auxiliary instruments were positioned even less visibly than in the standard model! This is a 1982-model four-door, with the new lockable centre cubby box.

With the four-door models launched, Land Rover next took a look at the Range Rover's transmission. The original LT95 four-speed gearbox with its integral transfer box had never been a very satisfactory unit. It had been designed in the days when the Range Rover had been conceived as an upmarket Land Rover, and while it was certainly tough, it was also pretty unrefined. Over the years, various improvements to the gearchange itself had made the Range Rover feel less agricultural on the road, but there was no doubt that the gearbox was a major hindrance to making the Range Rover a more refined road car.

During the 1970s, aftermarket conversions had been carried out in small numbers. Some had used automatic transmissions, initially the Borg Warner type 65 but later the Chrysler A727 Torqueflite which had been the very gearbox proposed by Spen King's New Vehicle Projects team at the end of the 1960s before production had started. Others had used the new five-speed Leyland LT77 gearbox, introduced in 1976 for the Rover SD1 saloon and later seen also in Triumph sports cars and Jaguar saloons.

Land Rover's customer research showed that there was a strong demand for an automatic transmission, and so priority was given to developing such an installation. The 1980 batch of twenty-five Schuler Automatic Range Rovers demonstrated that Land Rover were serious; the company told its overseas distributors at a September 1980 conference that an automatic model was on the way; and in *The Motor* of 11 July 1981 Managing Director Mike Hodgkinson admitted that there were 'several out in the field on test'.

Nevertheless, it was August 1982 before the Range Rover Automatic was announced. The chosen transmission was the Chrysler Torqueflite three-speed, which was mated to a two-speed transfer box, designed by Land Rover and designated the LT230R (LT stood for Leyland Transmissions; the R indicated that the intermediate gears were supported by roller bearings). On the whole, the installation was a resounding success, even though the Torqueflite was rather lazy and inefficient by later standards. Acceleration, top speed and fuel consumption were all worse than with the four-speed manual

Inside this 1983 four-door automatic can be seen the brushed velour seats with detachable cushions on their head restraints and the optional Rosewood door cappings. By this stage, the Range Rover was beginning to look like a luxury car.

transmission, but the Range Rover Automatic was very much easier to drive around town and suffered far less than the four-speed manual from transmission whine. It was also very much more controllable off-road than many commentators had expected, and this reflected Land Rover's determination not to give the vehicle better road manners at the expense of rough-terrain ability.

To help promote the new model, Land Rover announced a specially-equipped derivative which bore the cumbersome name of Range Rover Automatic In Vogue Special Edition. This came with the latest metallic paint colours, a unique coachline, and grey-painted alloy wheels – a styling feature shared with the contemporary Rover Vanden Plas saloon. There were special interior features, too, and the special edition also had all the features which were new for the 1983-season Range Rovers. Even without the air conditioning which could be ordered optionally, this was the most expensive Range Rover yet. Its price did not deter the buyers, though, and its ready market acceptance was a further indicator to Land Rover that their Range Rover development plans were right on target. During the 1983 calendar-year, Range Rover sales set new records in Britain and Europe.

It would probably have suited Land Rover to replace the old LT95 gearbox at a single stroke, but the production complications and cost implications of introducing two new transmissions at once were too great. So the four-speed carried on as the standard transmission for another year, giving way to the LT77 five-speed in July 1983. Once again, the choice of gearbox was no great surprise, and it came with the same LT230R transfer box as was fitted with the automatic transmission. The wand-like gear lever had hardly changed and the gearchange was still very far from slick, but the five-speed box was a great improvement. The overdrive fifth gear made the Range Rover quieter and more economical at cruising speeds, and there was a further reduction in transmission whine, although this was still a prominent aural characteristic of the latest Range Rovers.

Land Rover once again drew attention to their new transmission with a special-

The Automatic In Vogue Special Edition, 1982

Total built	Not known
Special features	Nevada Gold or Sierra Silver paint with broad twin coachlines (brown on Nevada Gold, grey on Sierra Silver)
	Three-spoke alloy wheels finished in grey enamel, with 'Range Rover' picked out in white on black centre caps
	Inlaid wood door cappings
	Cool box in rear loadspace
	Front and rear armrests
	Rear head restraints

This Special Edition was used to promote sales of the new Range Rover Automatic, and the base vehicle was a four-door with automatic transmission. Land Rover did not claim that it was a limited edition, and it may be that most home market four-door automatics sold in the 1983 season (that is 1982–1983) were Automatic In Vogue Special Edition models.

edition model. However, the third In Vogue variant, which arrived a month after the five-speed's introduction, was not just a showcase for the new gearbox. Automatics were now becoming so popular that Land Rover decided to offer a choice of manual or automatic transmission on this limited edition, which of course also featured the latest revisions and options for 1984-season four-door Range Rovers. As usual, it also had some unique features, as the specification panel below shows.

The 1983 Range Rover In Vogue

Total built	325 for UK market
Special features	Derwent Blue metallic paint with broad twin coachlines
	Three-spoke alloy wheels finished in grey enamel
	Walnut veneer door cappings
	Front and rear armrests
	Rear head restraints
	Rear seat belts
	Black tailgate capping
	Cool box in rear loadspace
	Picnic hamper in rear loadspace
	Digital radio/cassette with four speakers

The final In Vogue limited edition model was promoted in conjunction with the Daks autumn fashion collection at Simpson's of Piccadilly, London. It was a four-door model, and could be had with either automatic transmission or the new five-speed manual.

Key Production Changes 1979–1984

NB Paint and trim colour option changes are listed separately.

September 1979	Tape badges replaced plastic letters on bonnet and tailgate
	Scuttle side badges deleted
	Black polyester-coated bumpers replaced aluminium-painted type
	Repeater flashers added to front wings
	Rear lamp clusters incorporated fog-guard lamps
	Sundym tinted glass standardized
	Brushed nylon upholstery standardized
	Inertia-reel front safety belts standardized
	Head restraints standardized on front seats
	Four-spoke steering wheel replaced three-spoke type
	Warning lamps for rear fog-guard lamps and brake vacuum loss added to facia
	Power-assisted steering standardized
	Halogen headlamps standardized
	Uprated (25 ACR) alternator standardized
	Transfer box High Ratio raised to 1.113:1
	New option: Air conditioning
March 1980	Monteverdi four-door conversion announced (*see* separate panel)
June 1980	Chassis now electrophoretically corrosion-proofed before painting
	Narrower gearchange gate fitted

	Brushed velour upholstery (in Bronze) replaced brushed nylon
	Front seat head restraints now in velour with detachable cushions
	Rubber flooring and removable carpets added
	Carpet material added to rear wheel arches and lower door trim panels
	Fleet Line model introduced (*see* separate panel)
February 1981	In Vogue limited edition announced (*see* separate panel)
July 1981	Engine compression ratio raised to 9.35:1 (8.13:1 still available for export and to special order)
	Transfer box High Ratio raised to 0.996:1
	Intermittent wipe and wash/wipe cycle added to wash/wipe system
	Sun visors fitted with ticket pocket on driver's side and vanity mirror on passenger's side
	Courtesy lamp delay system fitted
	Map pockets added to front seat backs
	Carpet material replaces black material for spare wheel cover and tool kit curtain
	Radio aerial, front door speakers and wiring fitted as standard (except Fleetline models)
	Twin underbonnet lamps fitted (except Fleetline models)
	Four-door models announced
	Changed option: Stowage box fitted on transmission tunnel of two-door models with optional air conditioning
	New options (four-door models only): Electric windows Rosewood door cappings Loadspace carpet Three-spoke alloy wheels (All above available together with metallic paint as an Option Pack)
August 1982	New exhaust with burred-over tailpipe ends
	Overriders added to front bumper ends
	Coin tray replaced speaker grille on dash
	Leather covering added to steering wheel rim
	Column stalks changed to ISO standard (indicators now on the left, wash/wipe on the right)
	Rear centre armrest added (four-door models only)
	Detachable rear head restraints added (four-door models only)
	Adjustable inboard armrests added to front seats (four-door models only)
	New option: Automatic transmission (with transmission tunnel console, transmission oil temperature warning light on dash, transmission oil cooler ahead of radiator, and separate LT230R transfer box)
	Automatic In Vogue Special Edition announced (*see* separate panel)
July 1983	Five-speed LT77 gearbox and LT230R transfer box replaced four-speed LT95 gearbox with integral transfer box

	Transmission tunnel console fitted with five-speed gearbox Lower tailgate fitted with torsion bar balancing system Hydraulic jack replaced screw-type in tool kit Ribbed handbrake grip replaced smooth type All window pillars on two-door models painted satin-finish black Central locking fitted to four-door models (keylocks moved from door panels to handles) New options (for four-door models only): Option Pack C – front and rear armrests, rear head restraints, wood veneer door cappings, carpet for loadspace and lower tailgate Option Pack B – as Pack C, plus alloy wheels and metallic paint Option Pack A – as Pack B, plus air conditioning
August 1983	Third In Vogue Special Edition announced

Paint and Trim Options, 1979–1984

October 1979 to September 1980

There were eight standard paint options. Seats were upholstered in Bronze brushed nylon and the front head restraints were upholstered in Palomino PVC. The eight standard paints were:

Arctic White	Sahara Dust
Lincoln Green	Sandglow
Masai Red	Tuscan Blue
Russet Brown	Warwick Green

October 1980 to June 1981

There were nine paint colours, of which eight were standard options and one was available only on the special-edition In Vogue and on four-door models. This option (Vogue Blue) was the first metallic paint available on a factory-built Range Rover. Upholstery was in Bronze brushed velour, and the matching front head restraints had detachable cushions. The nine paint colours were:

Arctic White	Sandglow
Lincoln Green	Tuscan Blue
Masai Red	Vogue Blue
Russet Brown	Warwick Green
Sahara Dust	

July 1981 to September 1982

There were nine standard paint colours, of which the two metallics (Silver Birch and Vogue Blue) were available only on four-door models. Upholstery was in Bronze brushed velour, and the matching front head restraints had detachable cushions. The nine paint colours were:

Arctic White	Silver Birch
Masai Red	Tuscan Blue
Russet Brown	Venetian Red
Sahara Dust	Vogue Blue
Shetland Beige	

(Lincoln Green was also available to special order.)

October 1982 to September 1983
There were eight standard paint colours, of which the three metallics (Nevada Gold, Sierra Silver and Vogue Blue) were available only on four-door models. Upholstery was in Bronze brushed velour, and the matching front head restraints had detachable cushions. The eight paint colours were:

Arctic White	Shetland Beige
Nevada Gold	Sierra Silver
Russet Brown	Venetian Red
Sahara Dust	Vogue Blue

(Lincoln Green was also available to special order.)

October 1983 to May 1984
There were seven standard paint colours, of which the three metallics (Derwent Blue, Nevada Gold and Sierra Silver) were available only on four-door models. Upholstery was in Bronze brushed velour, and the matching head restraints had detachable cushions. The seven paint colours were:

Arctic White	Sahara Dust
Derwent Blue	Sierra Silver
Nevada Gold	Venetian Red
Russet Brown	

Chassis Numbers and Production Figures, 1979–1984

These figures show the build date, by calendar year, of Range Rovers built between 1979 and 1984. They were calculated by Anders Clausager of the British Motor Industry Heritage Trust from the factory Despatch Records which the Trust holds. They are used by kind permission.

From approximately October 1979, Range Rover chassis had standardized VIN (Vehicle Identification Number) codes. These consist of eight-digit (before 1982) or eleven-digit (1982 onwards) identifying prefixes followed by six-figure serial numbers. A typical Range Rover VIN would be SALLHAMV1AA105110. That number breaks down into ten groups, which can be decoded as follows:

1. SA	World manufacturer code for British Leyland (not found on pre-1982 models)	
2. L	Manufacturing sub-division (Land Rover Ltd; not found on pre-1982 models)	
3. LH	Model code (Range Rover)	
4. A	First type code	A = 100-inch wheelbase
5. M	Second type code	B = two-door M = four-door R = Monteverdi four-door
6. V	Engine type	V = 3.5-litre carburettor V8
7. 1	Steering and transmission code	1 = RHD, 4-speed manual 2 = LHD, 4-speed manual 3 = RHD, automatic 4 = LHD, automatic 7 = RHD, 5-speed manual 8 = LHD, 5-speed manual
8. A	Model-year code	A = all models to 1984

9. A	Assembly location	A = Solihull F = overseas (shipped as CKD)
10. 105110	Serial number	

VIN serial numbers started at 100001. Build dates were as follows:

1979	100001 to 102163
1980	102164 to 110584
1981	110585 to 119702
1982	119703 to 131429
1983	131430 to 143040
1984	143041 to 154589

NB The final figure given for 1984 is the final vehicle built during that calendar year. It is not possible to determine a final serial number for the 1984 model-year vehicles covered in this chapter, which ceased production some seven months before the end of the calendar year.

The 100 000th Range Rover was built during 1982.

4 Moving on up (1984–1989)

The summer of 1984 marked a decisive turning point for the Range Rover. Until then, luxury features had been largely limited to special-edition models, as Land Rover cautiously sounded out how much luxury the market would take in a dual-purpose vehicle. But by the summer of 1984 it was clear that the market would take all it was being offered and more – and so Land Rover introduced their first regular-production luxury model, which they called the Range Rover Vogue.

The first regular-production Vogue really represented the final stages of Mike Hodgkinson's five-year plan for the company; but Hodgkinson himself had already left Land Rover at the end of 1982. His place was taken by Tony Gilroy, a plain-speaking Irishman who had formerly been in charge of the Freight Rover light commercials operation under British Leyland. It was Gilroy who oversaw the final stage of the plans which Hodgkinson had approved for the Range Rover, and it was he who took the Range Rover into the next and even more successful phase of its existence.

Yet the company which Gilroy inherited was on a downward trend. Although the

The first regular-production Vogue models were announced in June 1984 and were distinguished from lesser types by their shaded alloy wheels and coloured bodyside tapes. However, from now on all four-door models would dispense with the quarter-lights in the front doors and would have the new mirrors seen here.

Range Rover was becoming ever more successful, that success was being more than offset by problems with the Land Rover utilities. Their sales overseas were collapsing as Japanese light utilities took more and more of their market, and it was obvious that something drastic had to be done. So it was that in 1983, Tony Gilroy commissioned a ground-up review of Land Rover products and markets. The full review took some two years to complete, but among its early recommendations were several which affected the Range Rover.

By this stage, the market for family 4x4 estates just below the Range Rover was being contested hard by Japanese manufacturers, and Land Rover knew that they were not yet in a position to fight back and win. It was therefore important to move the Range Rover away from this area of the market in order to protect its future. The obvious thing to do was to move it further up-market, and both engineers and marketing men believed that they could in time make it a credible alternative to the leading luxury saloons of the day in all respects.

With a more luxurious specification, the vehicle stood a good chance of success in the burgeoning sport-utility market in the USA, and so a further recommendation was that it should at long last be launched into the North American market for which it had originally been developed. Lastly, it would be important not to lose any of the vehicle's existing strengths in this redevelopment exercise, and so Land Rover committed themselves to maintaining the Range Rover's supremacy off-road.

Meanwhile, Gilroy had embarked on a ruthless drive to save money by streamlining Land Rover's production operations. November 1983 saw the first moves in a scheme to close down no fewer than nine 'feeder' factories in the Midlands and Wales and to bring their manufacturing and machining operations to the Solihull site. Space had been available since Rover car assembly had been transferred away from Solihull at the end of 1981, and it now made good sense to have both manufacturing and assembly on the same site. The changeover took some two years to complete, and was complete by the middle of 1985. The savings effected (around £14 million a year) helped to finance the launch of the Range Rover in the USA.

So it was that the next period of the Range Rover's development saw the vehicle become more and more luxurious, saw it enter the US market, and saw it consolidate its position as the best of the large four-wheel-drive estates. The period when Mike Hodgkinson had been Land Rover's Managing Director had already seen Land Rover switch from an engineering-led company to a market-led one; and the period of Tony Gilroy's stewardship would see that approach intensified to remarkable effect. The 1987 sales figures were more than 40 per cent higher than those for 1986; the 1988 figures were up again, to more than double their 1984 level; and the record year of 1989 saw more than 28,000 new Range Rovers find buyers. This was, admittedly, a time when Western economies were booming, and it could not last for ever. But Land Rover now had the right product to make the most of that boom.

THE MARKETING

Land Rover took great care over the marketing of the Range Rover at the beginning of this period, creating an image for the vehicle which would endure for the next decade. Promotional activities appealed to the individual streak in potential customers, as exemplified by a special promotion run in the UK between 10 September and 31 October 1984. Customers who took a test drive in a new Range Rover were presented with a pair of red Michelin travel guides in a special green slip-case, which was personalized with their name and bore the current Range Rover promotional legend: 'Luxury need not stop where adventure begins.'

By the mid-1980s, the Range Rover had become a status symbol, and promotional literature capitalized on this unashamedly. The 1986-season sales catalogue verged on the embarrassing in its decription of the vehicle. 'Designed for both professional and leisure pursuits,' it read, 'it will literally put you above the rest... A perceptive choice in transport; you will be forgiven for looking down on other motorists.' Only the most naïve of readers could have imagined that those words simply described the high-set driving position. And just in case the message was not clear, the brochure spelled it out: 'Few vehicles share such a diverse profile of regular buyers. It includes royalty, sports people, the armed forces, diplomats, police forces, industrialists – in fact, anyone with an important job to do'.

Things continued in much the same vein over the next few years. The 1987-season sales catalogue claimed that 'Range Rover owners do have a great deal in common. They know that the Range Rover is acceptable in every setting ... It is a load-carrier par excellence. But, for many owners, the ability to accommodate the Board of Directors will be justification enough to purchase it.' Like its owners, the Range Rover was established: 'Its classic good looks staunchly resist the vagaries of passing fashion'; and it was able to shrug off all competition: the Range Rover, 'despite the best efforts of its imitators, remains inimitable.'

With the arrival of the even more luxurious Vogue SE model in 1988, promotional material went a step further even than this. 'The best is now better than ever,' it read. 'And even more luxurious. Difficult to believe? Only until you take a look at the lavish new specially-equipped model.' This was a vehicle 'with all the class of a limousine'.

THE CUSTOMERS

Not all the Range Rover's customers in this period were the royalty, company directors and so on that promotional literature suggested, but there was no doubt that the vehicle had become a favourite of the professional classes. It undoubtedly did make a good tow vehicle and it undoubtedly had plenty of space for a family's camping gear, but most customers who wanted a vehicle for those purposes were now turning to the cheaper Japanese four-wheel drives: the Range Rover had become too expensive for them.

However, this was all part of Land Rover's plan. The Range Rover had to move into the luxury class so that the forthcoming new Discovery could take over the customers it had left behind. From the autumn of 1989, that was exactly what happened.

THE PRODUCT, 1984–1989

The luxury options available on four-door models were proving so popular, and customer reactions to the limited-edition In Vogue models had been so positive, that Land Rover judged the time was right to make a luxury edition Range Rover permanently available. So for 1985 there were high-line and low-line models. In most markets, the high-line model was called a Range Rover Vogue (although Australia knew it as the Range Rover Highline), while the low-line model remained a plain Range Rover wherever it was sold.

These new models were announced in Britain on 3 June 1984, and their main focus was on a very much improved interior. This was a by-product of the development work done to produce the more luxurious Vogue, the other changes made for 1985 being altogether less significant. In fact, the only mechanical change of any importance was the addition of Lucas electronic ignition, which helped the engine to remain in tune for longer periods and thus reduced maintenance requirements.

Nevertheless, there were other changes for the 1985 Range Rovers. Immediately obvious to anyone looking at the revised vehicles were new bonnet and tailgate badges, while the side view had also been tidied up by deleting the front quarterlights and integrating new rear-view mirrors with the leading lower corners of the window aperture. Their design, according to Land Rover stylist Tony Poole, was borrowed from a contemporary BMW saloon!

Under the skin, the 1985-model Range Rovers also had a new type of wiring harness, with thin-wall cables which allowed more wires to be incorporated with less bulk, and new multi-connector plugs were added to the harness to make the removal (or addition) of individual electrical components easier. For 1985, however, the Range Rover made only limited use of this new development; the real increase in its electrical convenience equipment was yet to come.

Those important interior changes included new Bronze Check cloth upholstery on the low-line models, more supportive front seats with height and tilt adjustment for the cushion together with reclining backrests, and upper belt mountings now fixed conventionally to the B-pillars on four-door models. Two-door models however retained their original seats with integrally-mounted belts. Door trims were also new on the four-doors, and incorporated plastic pulls taken from the Austin Maestro and Montego range. Loudspeakers, when fitted,

The 1985-model Vogue had new door trims, a new instrument pack, new grey seat fabric and a passenger-side grab handle, among other improvements.

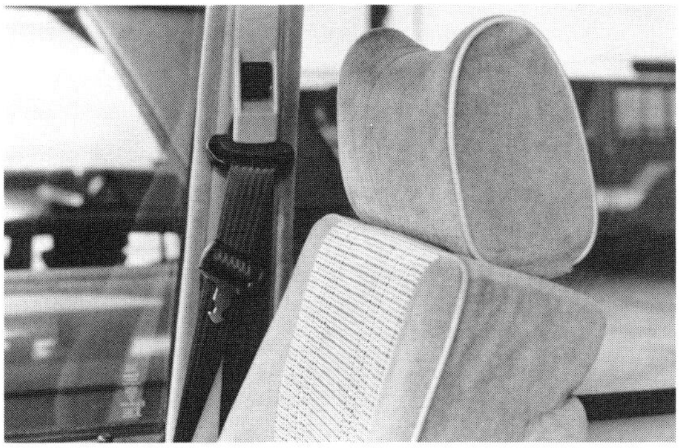

Another key change for the 1985 four-door models was in the seatbelts, which now had an adjustable upper mounting on the pillar. The head restraints were new, too. This picture shows the checked pattern of the standard Bronze Check upholstery.

were now mounted higher up the doors to give better audio quality.

A great deal had also been changed on the dash to improve its appearance. A redesigned and much larger instrument binnacle incorporated a rev counter for the first time, together with a bank of warning lights between its two main dials. The new instrument pack was borrowed – like the interior door handles – from the cheapest version of the contemporary Maestro, though few customers noticed and even fewer complained. The clock was now more visibly mounted in the middle of the facia and there was a new panel for minor switches alongside the heater control panel – which still stood uncomfortably proud of its moulded plastic surround. The passenger gained a grab handle and an open storage shelf in place of the old glove-box, although this shelf disappeared if air conditioning was fitted.

The new dashboard concealed a more powerful heater, and this fed both side window demister vents in the edges of the upper dash and (on four-door models) ducting which led to the rear passenger area. In the upper dash, the eyeball vents had gone, and in their place the main facia rail incorporated two rectangular vents, similar to those used with air conditioning.

The Vogue models announced in June 1984 of course incorporated all these refinements, and several others besides. Instead of Bronze Check upholstery, they had tasteful Silver Grey – the first deviation from brown on a Solihull-built Range Rover. A four-speaker system (with the rear speakers in the headlining above the tailgate) came with a state-of-the-art radio/cassette player, and the new door mirrors came with electric demisting and power adjustment. The contents of the rear loadspace were also concealed from prying eyes by a rigid shelf, which folded centrally across its width for ease of access. Unfortunately, it was positioned a little too low, so that suitcases could not be stood on end beneath it; it took Land Rover a further eight years to get that right!

The Vogue was distinctive outside, too, with grey-enamelled three-spoke alloy wheels and bodyside tapes which ran along its lower swage line. Sharp eyes would also have spotted the headlamp washer jets in the front bumper, and many Vogues were finished in the metallic colours which were available only on four-door models. As late as February 1984, the stylists had contemplated painting the radiator grille to match the body, as well, but this was not adopted for production.

Over the next two years, the main thrust of Range Rover development was to prepare the vehicle for its launch into the demanding US market, and the first of the production changes directed to that end became apparent on the 1986 models which were announced in October 1985. Land Rover were very much aware that they needed to improve their build quality, and so the 1986-model Range Rovers had improved paint and more consistent panel gaps, the latter resulting from the greater dimensional accuracy of jig-welded body-frames which was simply not possible with the original bolted frames.

Land Rover also recognized that it would be important to make the Range Rover feel more like a car on the road, and to that end they introduced a whole range of measures for the 1986 model-year. Body roll was reduced (though not enough) by some minor suspension revisions. Inside, a completely redesigned lower dash area provided a neater link to the transmission tunnel console and placed the radio and auxiliary switches more accessibly. The handbrake was relocated on top of the transmission tunnel, where a car driver would expect to find it, and a new direct-acting short-stick gearlever for the five-speed gearbox also gave the vehicle a more car-like feel. Meanwhile, the sluggish three-speed automatic transmission was replaced by a slick four-speed type, made by ZF in Germany, which had a lock-up clutch on its overdrive top gear.

The 1986 Vogue models were really the trailblazers for the US-market models, as they incorporated all these revisions and

An important change for 1986 was the arrival of the short-stick gear shift for manual models. The centre console was tidied up at the same time.

For 1986, the Vogue was distinguished by a bib spoiler with twin auxiliary lamps and by a rubbing-strip along its flanks. These were the first Range Rovers to have fuel injection. All models now had the wraparound bumper end caps.

more. They were characterized by a new front spoiler with twin driving lamps, plastic bumper end-caps and rubber side bump-strips with a bright insert, all of which gave them a more distinguished appearance. However, the real excitement of the 1986 Vogues lay under their bonnets in the guise of a fuel-injected engine. This bumped engine power up to 165bhp from the 127bhp of the carburetted engine, while placing the torque peak at higher revs so that road acceleration was improved.

The Lucas L-Jetronic fuel injection system was not new to the Rover V8, having been seen originally at the end of the 1970s on Rover 3500 saloons for Australia and the USA, and later in the Rover Vitesse and Vanden Plas EFi saloons. One reason for switching from carburettors was that fuel injection could provide more power without increasing the cubic capacity of the engine (as it had done in the later saloon engines), but another was that it allowed more precise tuning of the engine to meet emissions control regulations (as it had in the Australian and American saloons). For the Range Rover, Solihull wanted a compromise between the two. Most important was the development of an engine which would meet exhaust emissions regulations in the USA, but it was also essential to keep Range Rover road performance competitive with that of similarly-priced saloons in all markets.

That year, the Vogue accounted for more than 80 per cent of Range Rover sales in the UK, so relatively few home market customers would have realized that the carburettor engines had also been revised. In fact, the changes were quite remarkable, taking power up from 125bhp to 127bhp and torque up from 185lb/ft to 194lb/ft while improving control of exhaust emissions. The secret was a new Pulsair air injection system, which drew air into the exhaust ports through non-return valves during the negative pulse of the combustion cycle, thus allowing fuel to be burned more completely to the benefit of responsiveness and economy.

Nevertheless, one of the major hindrances to greater Range Rover sales in Europe continued to be the size and thirst of its engine. No matter that the V8 was superbly refined and powerful and that the Range Rover derived much of its essential character from it: in countries where engines bigger than about 2.5 litres were heavily taxed, it was not going to sell well. Similarly, there was widespread acceptance of diesel power for large 4x4 estates both in Europe and in countries like Australia, where Toyota's crude but rugged Land Cruiser was a best-seller in diesel form. As early as 1980, Land Rover had acknowledged that they would have to develop a diesel Range Rover, and they told their overseas dealers so at a conference in September that year.

The first plan for a diesel Range Rover was perhaps the most ambitious. Land Rover hoped to use the 'Iceberg' engine, a turbocharged diesel version of the petrol V8 which had been under joint development with Perkins Engines of Peterborough since the beginning of 1980. That engine promised 125bhp and 205lb/ft of torque when it was displayed in prototype form by Perkins at the Paris Motor Show in October 1982, but it never did reach production. The reasons are not clear, but it seems probable that Gilroy's review of Land Rover had shown the likely demand for this engine in all its applications was insufficient to justify the cost of tooling-up for production.

As there was no other suitable diesel engine being made or under development within British Leyland, Land Rover had to buy one in. In fact, they had started to look

Range Rover 1984–1989

Layout and chassis

Two-door or four-door estate with horizontally split tailgate
Box-section ladder-frame chassis

Engine (petrol)

Type	Light-alloy 90-degree petrol V8
Capacity	3,528cc (3,947cc for USA only, 1988–9)
Bore and stroke	88.9mm x 71.1mm (93.98mm x 71.1mm for 3.9-litre type)
Fuel system	Two Zenith-Stromberg carburettors (1984–6)
	Lucas L-Jetronic fuel injection (1985–6 Vogue and all subsequent models)
Compression ratio	9.35:1 (8.13:1 for all US-market engines)
Max. power	125bhp @ 4,000rpm (carburettor engines)
	165bhp @ 4,000rpm (injected engines, except for USA)
	150bhp @ 4,000rpm (injected 3.5-litre for USA)
	179bhp @ 4,750rpm (US-market 3.9-litre)
Max. torque	185lb/ft @ 2,500rpm (carburettor engines)
	206lb/ft @ 3,200rpm (injected engines, except for USA)
	190lb/ft @ 3,200rpm (injected 3.5-litre for USA)
	227lb/ft @ 3,500rpm (US-market 3.9-litre)

Engine (VM turbodiesel)

Type	VM type HR 492 HI OHV in-line four-cylinder turbocharged and intercooled indirect-injection diesel
Capacity	2,393cc
Bore and stroke	92mm x 90mm
Compression ratio	22:1
Max. power	112bhp @ 4,200rpm
Max. torque	183lb/ft @ 2,400rpm

Transmission

Primary gearbox	Five-speed LT77 manual (1984–9)
	Optional three-speed Chrysler A727 automatic (1984–5)
	Optional four-speed ZF 4HP22 (1985–9)
Ratios	Five-speed: 3.32:1, 2.13:1, 1.39:1, 1.00:1, 0.77:1, reverse 3.42:1
	Three-speed automatic: 2.45:1, 1.45:1, 1.00:1, reverse 2.20:1
	Four-speed automatic: 2.47:1. 1.47:1, 1.00:1, 0.72:1, reverse 2.08:1
Transfer gearbox	Two-speed LT230R (1984–5) or LT230T (1985–8); two-speed Borg Warner (1988–9)
Ratios	High 1.19:1, or 1.003:1 (1984-5 with three-speed automatic primary gearbox)
	Low 3.32:1
Centre differential lockable	
Final drive	3.54:1

Suspension and steering

Front	Live axle with coil springs and hydraulic telescopic dampers, located by radius arms and Panhard rod
Rear	Live axle with coil springs and hydraulic telescopic dampers, located by radius arms, support rods and central wishbone assembly incorporating a Boge Hydromat self-energizing ride-levelling strut
Steering	Power-assisted recirculating-ball, worm and nut
Ratio	17.5:1
Tyres	205 x 16
Wheels	Pressed steel 16in Rostyle type; three-spoke alloy type on some models

Brakes

Type	Disc brakes with dual hydraulic line and servo assistance; four-piston callipers on front wheels and two-piston type on rears; internal expanding drum-type parking brake operating on transfer box output shaft
Size	Front 11.75in diameter; rear 11.42in

Vehicle dimensions (in/mm)

Wheelbase	100/2,540
Overall length	176/4,470
Overall width	70/1,780
Overall height	70/1,780
Track	58.5/1,480
Kerb weight	Two-door models 3,872lb/1,756kg Four-door models 3,942lb/1,788kg

Performance

Max speed	102mph/164km/h (with injected 3.5-litre engine) 92mph/148km/h (with turbodiesel engine)
0-60mph	12 sec (with injected 3.5-litre engine) 16.5 sec (with turbodiesel engine)

at alternative power units even before the Iceberg V8 was cancelled, and among those they considered was the Peugeot XDT 2.5-litre four-cylinder (possibly in XDTS turbocharged form). Rumour also has it that they tried the brand-new BMW 2.4-litre six-cylinder, but the engine which found favour was one made by VM Motori in Italy. After trying out a five-cylinder 3-litre turbodiesel, Land Rover settled on an intercooled version of the turbocharged four-cylinder 2.4-litre engine which VM were already supplying to Rover for the 2400SD Turbo saloon. The lower-powered, non-intercooled saloon version of the engine in fact already enjoyed Land Rover Approval as an aftermarket conversion by Janspeed of Salisbury.

Land Rover announced the Range Rover Turbo D at the Turin Motor Show towards the end of April 1986, the choice of an Italian show being dictated partly by the engine's Italian origin and partly by Land Rover's belief that sales of the new model would be strong in Italy. Meanwhile, pre-production examples (not quite standard,

After the fuel-injected version of the V8 came a turbodiesel model, with a four-cylinder engine built by the Italian VM company.

despite claims to the contrary) were used to gather publicity in the 'Project Bullet' record runs at the MIRA proving ground near Nuneaton. One of the Bullet Range Rovers became the first diesel vehicle to run for 24 hours at an average of over 100mph (160km/h) (a speed which production models could not even reach), and the two set no fewer than twenty-seven new British speed records for diesel vehicles.

Although the turbodiesel engine's 112bhp kept maximum speed below that of the more powerful petrol models, its 183lb/ft of torque meant that acceleration was as good as that of a carburetted Range Rover. Of course, no four-cylinder engine can be as smooth as an eight-cylinder type, and the VM turbodiesel was a little rough; diesel knock was also a drawback, although it had been tamed by good soundproofing; and the diesel was available only with a five-speed manual gearbox and with the basic two-door or four-door specification. These factors were enough to deter buyers in diesel-shy Britain, where the Turbo D was never a strong seller, but in continental Europe it soon began to outsell the petrol versions by a considerable margin. In due course, 40 per cent of all Range

The record-breaking Project Bullet vehicle pictured at speed. It is now preserved in the Heritage Collection at Gaydon.

The Range Rover Turbo D was distinguished by special tailgate badging. It was available only with the base specification and thus did not have a spoiler. Just visible in this picture is the single-outlet exhaust pipe; V8s still had a dual-outlet exhaust.

Rovers built would be diesel-powered.

The turbodiesel Range Rover was of course of no interest whatsoever to the US market, and so it was left out of the reckoning when Land Rover came to make the final series of revisions for the US launch. The US-model Range Rover was developed under the code name of Eagle, and although it would not be announced until March 1987, many of the features developed specially for it became available on Range Rovers for other markets in December 1986; in fact, Henly's showroom in London (and probably others) had these new vehicles by the end of November, before the official launch date.

In Land Rover's terms, these were the 1987-model Range Rovers. Their most obvious distinguishing features were a new grille with horizontal slats and a flap which concealed the fuel filler and thus tidied up the side appearance. Concealed bonnet hinges also helped the tidier look, and the bonnet was now easier to lift thanks to counterbalancing torsion bars. It was less of a chore to open the lower tailgate, too, as its release handle had been relocated inside so that owners would no longer get their hands dirty on the original crude external release lever.

Inside, a split rear seat offered greater loadspace versatility, and the front seats had been moved closer to the centre of the car to make room for map pockets in the door trims. A smaller steering wheel with two spokes contributed to the car-like feel of the interior and also made the steering feel more positive, while new stalk switches (once again sourced from the Maestro and Montego) and air conditioning controls

For 1987 came a grille with horizontal slats and the indicator repeaters were repositioned lower down. The bonnet now had concealed hinges and the filler cap was covered by a door – exactly as Rover had intended twenty years ealier!

integrated with the heater controls also made the vehicle feel more like a car and less like an upgraded truck. With all these features – and with one or two which were exclusive to the Eagle project – the Range Rover was ready for its launch on the US market.

Over the next two years, Solihull devoted a great deal of effort to making the Range Rover a success in the USA, backed up by a first-rate team at Range Rover of North America. The first year's sales target of 3,000 vehicles was exceeded by nearly 10 per cent, despite a slump in the sales of imported cars generally and the stock market crash of October 1987. Listening carefully to what the US customers wanted, Land Rover made some adjustments for the 1988 models: in particular, the demand for white Range Rovers and the demand for

The Range Rover Olympic

Land Rover were concerned about the success of Japanese four-wheel drive estates, and in 1986 they started work on a directly competitive vehicle which would materialize in 1989 as the Discovery. In the meantime, however, they looked at the possibility of building a more sporting Range Rover to capture some of the market the Japanese had created.

Three such vehicles were built. The first one was displayed at the 1986 Motor Show, partly to deflect attention from the fact that Land Rover were not announcing any changes for the Range Rover (the 1987 models appeared a couple of months later, at the end of the year). Birmingham was then in the running as the host city for the next Olympic Games, so Land Rover capitalized on this by calling their vehicle the Range Rover Olympic.

The Olympic had a two-door body with Recaro sports seats, the latest four-door spoiler, and painted three-spoke alloy wheels. It was finished in gunmetal grey with red 'Olympic' decals on the doors. A second vehicle was produced to a similar specification, this time painted red. The third and final example was built in 1987 and was registered as E 970 EAC. This one was finished in metallic gunmetal, with red highlights on its side bump-strips. It had one-piece rear side windows and a prototype 3.9-litre engine.

Land Rover decided against marketing a sporting two-door model of the Range Rover, but the idea was revived in 1990 for the limited-edition Range Rover CSK.

The Range Rover Olympic prototypes were built to examine ways of creating a sporting two-door model. This is the third one, built in 1988. No production two-door model had either the spoiler or the three-spoke alloy wheels at the time E 970 KAC was built; inside were Recaro sports seats and under the bonnet was a prototype 3.9-litre V8 engine.

the optional leather upholstery had exceeded expectations, and so production was adjusted accordingly. However, the only really new features for 1988 were cruise control and a slide-and-tilt sunroof – and that sunroof was the only really significant change introduced on the 1988-model Range Rovers for the rest of the world in October 1987.

Equipment levels on US-model Range Rovers would remain higher than those on other models for some time to come, and the panel on page 73 shows how some features introduced for the US market would take several years to filter down to Range Rovers for other markets. One reason for this was that Solihull was not sure whether the rest of the world wanted such high equipment levels, or – perhaps more important – whether it was prepared to pay for them. So Land Rover proceeded with caution, just as they had done with the In Vogue and other special editions in the early 1980s. To test the water, they pulled together a number of the US-market features on a model they called the Vogue SE, and announced it at the Geneva Motor Show in March 1988.

The SE of the name stood for Special Equipment, and those letters precisely summarized what this new vehicle was about. It came only with metallic paint, with sunroof, air conditioning and automatic transmission, and it had the leather upholstery which was optional in the USA together with electrically-adjustable front seats. Burr walnut wood trim was a new

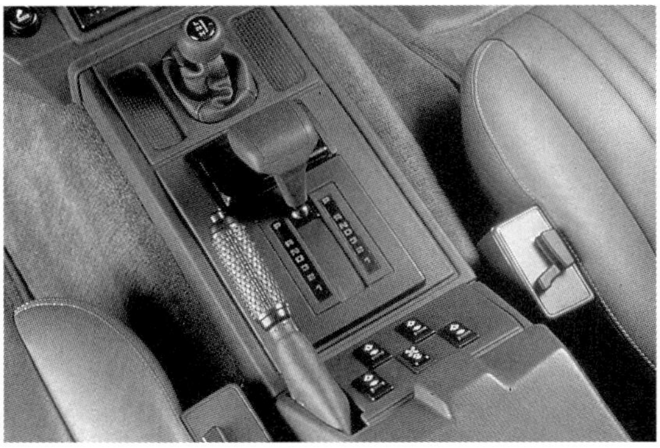

The automatic shift lever had looked like this ever since the ZF box arrived in 1986, but the single-plane transfer box lever marks this out as a 1989 model. It is actually a US-specification Range Rover, and those boxes alongside the leather seats house the controls for the seats' power adjustment.

The big news for 1988 was an electric sunroof option, seen here on a UK-specification Vogue model.

This 1989-model Vogue SE shows off the latest spoiler with four slots in its lower edge, together with the body-coloured wheel and the pinstripe coachline which were then peculiar to the SE.

feature, and from the outside, the SE was distinguished by alloy wheel spokes painted the same colour as the bodywork and a white pinstripe coachline on the sides.

The Vogue SE was a rather bigger hit than Land Rover seem to have expected, and so it was retained in the line-up for 1989, further distinguished by tailgate badging and by additional wood trim inside, both in response to customer demand. And in fact there would be a Vogue SE right through to the end of Range Rover production, because it was exactly what the market wanted – a vehicle with all the towing and rough-terrain capabilities expected of the Range Rover allied to a no-holds-barred luxury specification. Over the next few years, Land Rover would use the Vogue SE flagship to push the Range Rover further and further up-market, while retaining the regular Vogue model as a cheaper but nonetheless well-equipped version. As for the base models, these had practically disappeared from most markets by the end of the 1980s, and in many countries (Britain included), the two-door

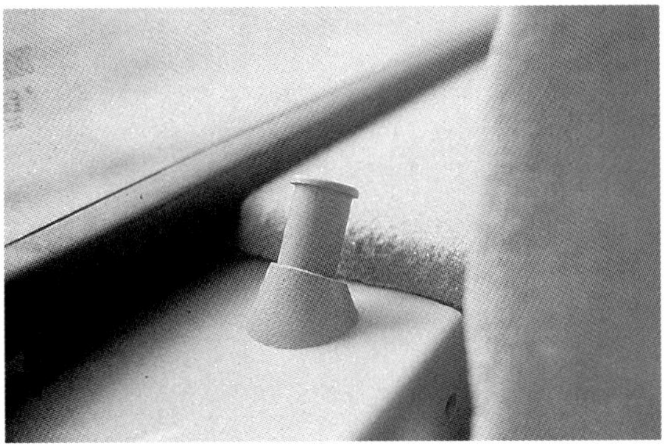

The 1989 models had this neat push-button release for the rear seat in place of the earlier pull-handle.

was now available only to special order.

The 1989-model Range Rovers continued very much in the vein of the 1988 models, but they did come with two valuable new refinements. Interiors were made more car-like with neat new door trims, and on SEs the burr walnut was extended to the facia rail and centre console. On the mechanical side, meanwhile, there was a new transfer box with a chain drive which completely eliminated the shrill whine of the gear-driven LT230T box. No doubt the US market had again inspired the change, as many US-built 4x4s used chain-driven boxes by Borg Warner, whose product the new Range Rover box was. The chain drive was accompanied by a viscous coupling which automatically locked the centre differential if it detected the wheels on one axle turning significantly faster than those on the other. This made off-road driving very much simpler by relieving the driver of the need to judge when to lock the centre differential, and it could also improve on-road traction in some conditions.

By the time this 1989-model Vogue was built, the interior looked much more integrated. This example has the standard cloth seats, but unusually has a manual transmission and no air conditioning.

For 1989, the Land Rover-built transfer box was replaced by a new type made by Borg Warner. This cutaway shows the silent chain drive which replaced the noisy gear drive of the older box, and the viscous coupling in the centre differential which superseded the manual lock.

The North American Range Rover, 1987–1989

Land Rover put a great deal of effort into selling the Range Rover in the USA, and they were rewarded with excellent sales and reports of high customer satisfaction. So important did the US market become to the Range Rover in the late 1980s that several features were introduced on US models before they became available elsewhere. Examples are leather upholstery and electrically adjustable front seats (March 1987 in the USA, March 1988 elsewhere), cruise control (October 1987 in the USA, October 1990 elsewhere), the tilt-and-slide sunroof (October 1987 in the USA, March 1988 elsewhere), and the 3.9-litre engine (October 1988 in the USA, October 1989 elsewhere).

Brief specification details of the 1987–1989 US model Range Rovers are as follows:

March to October 1987
Essentially the same as the contemporary UK Vogue. Air conditioning, six-way power adjustable front seats, ICE theft deterrent system and tow hitch plate all standard. Cloth upholstery standard, leather optional. 150bhp 3.5-litre V8 engine with catalytic converter; automatic transmission standard with no manual option.

October 1987 to September 1988
Transmission lock-up delayed to give improved response in town driving. Cruise control standard. New brake servo and new disc material to give better brake pedal feel. Radio aerial in rear window heating element. Revised air conditioning ducts. Four new paint colours. Leather upholstery more widely available. Optional electric slide-and-tilt sunroof.

October 1988 to September 1989
178bhp 3.9-litre V8 engine with catalytic converter; Borg Warner chain-driven transfer box; automatic transmission standard with no manual option. Eight-way power adjustable front seats. New door trims and six speaker ICE system. Tan interior trim available. Further improved ventilation and 20 per cent increase in air conditioning cooldown. Single-touch operation for driver's window. Tailgate included in central locking system. Heated windscreen and heated washer jets. Ignition headlamp override. One new paint colour.

Key Production Changes, 1984–1989

NB Paint and trim colour option changes are listed separately.

June 1984	Vogue introduced as regular production model
	Front quarter-lights deleted (four-door models only)
	Larger door mirrors, fitted to leading edge of front window apertures
	Colour-keyed front and rear tape badges replaced black-and-white type
	Weather strip added between lower tailgate and bumper
	Black tailgate capping
	Electronic ignition added to engine
	Height-adjustable reclining front seats replaced non-adjustable type (four-door models only)
	Height-adjustable safety-belt mountings on centre pillars (four-door models only)
	Bronze Check cloth trim replaced Bronze velour; Silver Grey cloth for Vogue models
	New front door trims with radio speakers built in
	New dashboard with passenger grab handle, side window demister vents, fuse panel, open parcel shelf instead of glove box (except with air conditioning)
	Heater with 50 per cent more output and with air bleed to rear passenger area
	Larger instrument panel incorporating rev counter
	Full centre console with cubby box added
	Intermittent wipe incorporated in rear wash-wipe
	Revised upper tailgate struts
	New type of electrical wiring harness
	New options (all standard on Vogue):
	Door mirrors with electric adjustment and electric demisting
	Headlamp washer jets
	Rear loadspace cover
	Four-speaker stereo radio/cassette system
October 1985	Fuel-injected engine (Vogue models only)
	Carburettor engine with 127bhp replaced 125bhp type
	'Short-stick' direct gearchange for five-speed manual models
	Four-speed ZF 4HP22 automatic gearbox replaced three-speed Chrysler type
	Revised rear suspension with two forward-facing dampers instead of staggered layout and with dual-rate coil springs
	Larger front radius arm bushes for improved roll stiffness
	All-welded bodyframe replaced bolt-together type
	Side rubbing strips with bright insert, polyurethane bumper end caps and front spoiler with twin quartz-halogen driving lamps (Vogue models only)
	Red warning lamps added to front door trailing edges and puddle lamps added to lower shut faces
	New centre console added, with integral stowage box, rear ashtray and rear cigarette lighter
	Backlit heater and air conditioning control panel added

	Radio relocated from driver's side of dash to centre console Bracken trim replaced Bronze Check (Vogue models only) New options: Avon Rangemaster 215/75x16 tyres (Vogue models only)
April 1986	Turbo D model announced, with 2.4-litre VM diesel engine and five-speed manual transmission LT230T transfer box (with taper bearings on intermediate gears) replaced LT230R (date approximate)
November 1986	Fuel-injected engine standardized on four-door models Bonnet fitted with concealed hinges and counterbalancing torsion bars Horizontally slatted plastic radiator grille replaced vertically-slatted metal type; lower edge now painted in body colour Plastic-bodied front and rear lamp units replaced metal-bodied type Side repeater lamps relocated from upper wing to above bumper wraparound Plastic flap added over fuel filler and connected to central locking system Lower tailgate release repositioned inside Fixed rear number plate replaced hinged type Black plastic rear number plate lamps replaced metal type Upper tailgate lock now used ignition key Redesigned rear window heating element, incorporating radio aerial Front seats relocated closer together Rear seat redesigned with integral head restraints and one-third/two-thirds split (four-door models only) Map pockets added to front door trims (four-door models only) Two-spoke steering wheel without leather rim cover replaced larger four-spoke type Redesigned steering column shroud, incorporating hazard warning lights switch Backlit, multi-function column stalk controls replaced simpler type Additional warning lamps for washer bottle and brake fluid levels Grab handle for front seat passenger added to cantrail Improved operation of safety belt reels Improved radio speakers fitted Electric door mirror controls moved to dashboard from column shroud Alloy wheels now with asymmetric rim hump to improve retention of deflated tyre Changed options: Wooden door cappings now without visible fixing screws (standard on Vogue models) Air conditioning controls now integrated with heater controls (standard on Vogue models)
October 1987	New roof panel without ribbing over front seat area New type of brake servo New option: Electric tilt-and-slide sunroof (Vogue models only)
March 1988	Vogue SE model announced
October 1988	Borg-Warner chain-driven transfer box with viscous-coupled centre differential replaced LT230T gear-driven type with manually lockable centre differential Land Rover badge added to grille

	Front spoiler with four slots in lower edge replaced earlier type with plain edge
	Variable delay system fitted for front wipers
	Heated front washer jets
	More complex rear wash/wipe programme
	Larger heated rear window element
	Ignition-off headlamp override
	Central locking extended to include tailgate
	Vogue SE decal badge added to tailgate
	(Vogue SE models only)
	Rear view mirror mounted to windscreen instead of roof
	Single-plane selecter lever for new transfer box and revised centre console with additional coin tray
	Leather-covered gear knob on manual models replaced black plastic type
	Lip added to lower edge of radio panel
	Single-touch electric window operation for driver's door (four-door models only)
	Lightweight (4mm) side glass replaced 5mm type
	Deflector fitted to leading edge of front window frame
	New door trims with integrally moulded armrests, new door release handles and flush-fit locking buttons
	Larger storage pockets on front door trims
	Polished American Walnut wood inserts in door trims
	Press-type releases for rear seat replaced pull-type
	Changed option: More powerful fan for air conditioning
	New option: Upgraded radio/cassette with additional bass speakers in front doors
May 1989 (approx)	Deeper underbody sills fitted
	Concealed front door hinges

Paint and Trim Options, 1984–1989

June 1984 to September 1985

There were nine standard paint options, of which the three metallics (Derwent Blue, Nevada Gold and Sierra Silver) were available only on four-door models. Vogue models had bodyside tape stripes in Brown, Green or Silver. Bronze Check cloth upholstery was standard; Vogue models had Silver Grey velour upholstery. The nine paint colours were:

Arctic White	Russet Brown
Arizona Tan	Sahara Dust
Balmoral Green	Sierra Silver
Derwent Blue	Venetian Red
Nevada Gold	

September 1985 to November 1986

There were 10 standard paint colours, of which the four metallics (Astral Silver, Caspian Blue, Cypress Green and Savannah Beige) were available only on four-door models. Vogue models had bodyside tape stripes. Bronze Check cloth upholstery was standard; Vogue models had Silver Grey velour upholstery. The ten paint colours were:

Arizona Tan	Chamonix White
Astral Silver	Cypress Green
Balmoral Green	Savannah Beige
Cambrian Grey	Tasman Blue
Caspian Blue	Venetian Red

December 1986 to September 1987

There were 10 standard paint colours, of which the five metallics (indicated by an 'm' on the list below) were available only on four-door models. There were three upholstery options: Bronze Check upholstery was standard, Vogue models had either Silver Grey or Bracken velour (see below), and Bracken velour was also provided when the Interior Pack was specified for a standard vehicle in any paint colour. The options were:

Ascot Green	Vogue models trimmed in	Bracken
Astral Silver (m)		Silver Grey
Cambrian Grey		Silver Grey
Caspian Blue (m)		Silver Grey
Cassis Red (m)		Silver Grey
Chamonix White		Silver Grey
Cypress Green (m)		Silver Grey
Savannah Beige (m)		Bracken
Tasman Blue		Silver Grey
Venetian Red		Bracken

October 1987 to September 1988

There were 11 standard paint colours, of which the six metallics (indicated by an 'm' on the list below) were available only on four-door models. Vogue SE models (from March 1988) were available in Caspian Blue and Cypress Green only. There were three upholstery options: Bronze Check upholstery was standard, Vogue models had either Silver Grey or Bracken velour (see below), and Bracken velour was also provided when the Interior Pack was specified for a standard vehicle in any paint colour. Vogue SE models had grey leather upholstery as standard. The options were:

Ascot Green	Vogue models trimmed in	Bracken
Alaskan Blue (m)		Silver Grey
Cambrian Grey	Vogue models trimmed in	Silver Grey
Caspian Blue (m)		Silver Grey
Cassis Red (m)		Silver Grey
Colorado Silver (m)		Silver Grey
Chamonix White (m)		Silver Grey
Cypress Green (m)		Silver Grey
Portofino Red		Silver Grey
Savannah Beige (m)		Bracken
Shire Blue		Silver Grey

October 1988 to September 1989

There were twelve standard paint colours, of which four were metallics (indicated on the list by 'm'), two were micatallics (indicated by 'mca') and one was a clear-over-base type (indicated by 'cob'). Vogue SE models were available in only four colours (Beluga, Cairngorm Brown, Caspian Blue and Cypress Green), and one colour (Beluga) was exclusive to the Vogue SE. Standard upholstery was in Winchester or Brogue cloth; Vogue SE models came with Winchester or Sorrell leather upholstery. Seat belts and interior plastic trim changed from blue-grey to Bokhara (brownish grey). The options were:

Alaskan Blue (m)	with trim in	Winchester
Arran Beige		Winchester
Beluga (cob)		Winchester
Cairngorm Brown (mca)		Brogue or Sorrell (SE)
Caspian Blue (m)		Winchester
Chamonix White		Brogue or Winchester
Colorado Silver (m)		Brogue or Winchester
Cypress Green (m)		Winchester
Eastnor Green		Brogue
Portofino Red		Winchester
Shire Blue		Winchester
Trocadero Red (mca)		Winchester

Chassis Numbers and Production Figures, 1984–1989

In this period, Range Rover chassis numbers continued with the system started in 1982 of an eleven-digit VIN code followed by a six-digit serial number.

A typical number would be SALLHAML7FA380728. This decodes as follows:

1. SA	World manufacturer code for Austin Rover (to 1986) or Rover Group (from 1986)	
2. L	Manufacturing sub-division (Land Rover Ltd)	
3. LH	Model code (Range Rover)	
4. A	First type code	A = 100-inch wheelbase
5. M	Second type code	B = two-door M = four-door
6. L	Engine type	E = 2.4-litre turbodiesel L = 3.5-litre injected V8 V = 3.5-litre carburettor V8
7. 7	Steering and transmission code	3 = RHD, automatic 4 = LHD, automatic 7 = RHD, 5-speed manual 8 = LHD, 5-speed manual
8. F	Model-year code	B = 1985 models C = 1986 models D = 1987 models E = 1988 models F = 1989 models
9. A	Assembly location	A = Solihull F = overseas (shipped as CKD)
10. 380728	Serial number	

Build dates were as follows:

1984	143041 to 154589
1985	154590 to 167941

Land Rover Ltd quote the following production figures for Range Rovers in the remaining period. These figures equate to calendar year rather than to model-year.

1986	14,494
1987	20,815
1988	24,021
1989	28,513

5 The Undisputed King (1989–1995)

The period between 1989 and the Range Rover's twenty-fifth production anniversary in 1995 was no less fascinating than the years which had gone before. Far from allowing the now elderly design to fade away gracefully, Land Rover continued to develop it and to strengthen its position in the luxury car market to prepare the way for the second-generation Range Rover which would be launched in 1994.

In this period, there were more major changes at Solihull, but the Range Rover was by now so well established that they made little difference to its progress. The first one came in 1988, when Tony Gilroy stepped down from the Managing Director's job at Land Rover and was not directly replaced. Instead, the senior man at Solihull became Chris Woodwark, the company's Commercial Director. He was replaced in turn four years later by John Russell. Even more far-reaching for Land Rover as a whole was the sale of its parent Rover Group to the German BMW company in January 1994. Yet none of this directly affected the Range Rover.

A much bigger impact on the way the Range Rover developed over the next few years was made by the introduction of the Land Rover Discovery in November 1989. The Discovery was of course based on a Range Rover chassis and powertrain, but it was aimed at a sector of the market just below the Range Rover – one where the Japanese had been predominant with family-oriented 4x4 estates and where Land Rover had been able to offer nothing except the rather utilitarian County Station Wagons. Inevitably, the Range Rover had to be kept one or two steps ahead of its new sibling, and Land Rover's marketing people carefully juggled specification levels between the two ranges to ensure that there were no damaging overlaps.

However, even though the Discovery's evolution depended to some extent on Range Rover hand-me-downs, elements of the younger vehicle's specification also filtered through to the Range Rover. Thus, it donated its new turbodiesel engine, its fuel tank, and its optional five-spoke alloy wheels – to name but three of the more obvious items. Marketing of the Discovery, with its vast range of accessories and options which enabled customers to individualize their vehicles, also taught Land Rover some new marketing tricks which were used to good effect on the Range Rover. In the early 1990s, there were several well-specified 'special edition' models, particularly in continental Europe, which drew on the lessons learned from Discovery marketing. These also helped to reinforce the aura of exclusivity which had always been part of the Range Rover image.

This was also a period in which Range Rover sales in the USA became even more important and in which US market demands heavily influenced the way the Range Rover was developed. It was a period in which the vehicle was introduced into some important new markets – Japan and Canada among them – and it was a period in which it successfully fought off serious challenges from newer vehicles like the 1989 Toyota Land Cruiser VX and the 1991 Mitsubishi Shogun to remain the undisputed king of the 4x4 luxury estate market. Sales did drop, partly as a result of the worldwide recession in 1991–2 but also because the Range Rover was deliberately being pushed into a market where cost inevitably limits sales volume. Yet Land Rover were ready to replace those lost Range Rover sales – down from over 28,000 in 1989 to under 18,000 in 1993 – with the top-model Discoverys.

THE MARKETING

For 1989, the main Range Rover sales catalogue opened with the statement that the vehicle was 'firmly established as one of the world's most admired and coveted vehicles'. That in fact was a true reflection of its position in the market, and promotional activities over the next six years were mainly directed towards maintaining that position.

By now, the Range Rover had been on sale for twenty years, and Land Rover celebrated the occasion briefly during 1990 with a few special press releases. However, there was no 20th Anniversary special edition, probably because the company was not keen to emphasize how old its flagship was when it was going to remain on sale for another five years or more!

Sales catalogues touched on its age discreetly, turning it into a virtue, and the 1992 model-year catalogue spoke of the vehicle's 'timeless style, emphasizing the difference between arriving and merely turning up. Yet it is a style derived from function: Range Rover projects a discreet confidence, a suggestion that when the going gets tough, any response will come in

With the 1990 models came both new upholstery material and a lateral stitch pattern for the centre panels. This is a four-door UK-market Vogue.

a manner utterly appropriate to the circumstances ... Range Rover is an enduring success story, proof positive that, in an increasingly identikit world, there remains a place for genuinely distinctive vehicles.'

By 1995, and the Range Rover's twenty-fifth year in production, there was almost nothing left to say. As that year's sales catalogue put it: 'all around the world, people have chosen the Range Rover, above all others, to provide them with the ability to travel everywhere, in luxury and comfort, safe in the knowledge that, when they've chosen Range Rover, they've arrived.' From this point on, the Range Rover was allowed to sell itself, and all the marketing effort was put behind the second-generation vehicle which was announced in September 1994.

THE CUSTOMERS

The Range Rover's target customers remained broadly the same for the first half of the 1990s as they had been at the end of the 1980s, with one important exception. The introduction of the long-wheelbase model in 1992 took the Range Rover into an even higher price bracket, and thus attracted a small percentage of new buyers for whom the original vehicle was too down-market.

The extra lounging room of the long-wheelbase model also took the vehicle into the chauffeur-drive market, although few examples seem to have been bought for this purpose at the time. Range Rover buyers on the whole remained owner-drivers, whether they bought a standard or long-wheelbase model, and this probably explains why a 'mobile office' conversion prepared for the long-wheelbase models by the Special Vehicles division at Solihull failed to arouse much interest.

THE PRODUCT, 1989–1995

So many detail changes were made to the Range Rover during this period that it is sometimes difficult to form a picture of the most important happenings. These began in October 1989 as the 3.9-litre petrol engine replaced the 3.5-litre type for all markets, while the 2.5-litre turbodiesel

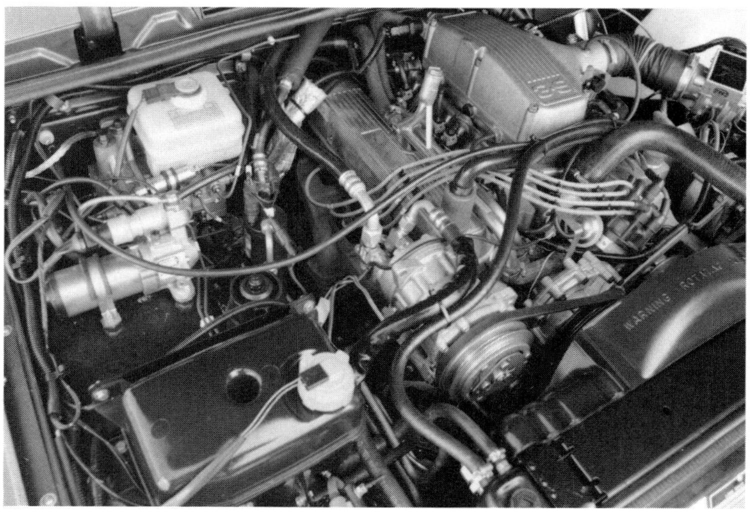

The big news for 1990 was that the 3.5-litre injected V8 had been replaced by the 3.9-litre type first seen in US-model Range Rovers a year earlier ...

... but many European markets were more impressed by the enlarged turbodiesel engine introduced at the same time. Still built in Italy by VM Motori, it now displaced 2.5 litres.

replaced the original 2.4-litre type. At the same time, ABS became an option on top models. A year later came anti-roll bars, concealed door hinges and a new fuel tank.

Things went quiet during 1991 – for reasons explained below – but October 1992 saw the introduction of the long-wheelbase Range Rover with its air suspension and traction aids. At the same time, Land Rover's own intercooled turbodiesel engine replaced the older VM diesel. In October 1993 came Autobiography, the scheme under which Range Rovers could be custom-built, and then in March 1994 came a new dashboard with airbags, plus a new manual gearbox. From October that year, with the introduction of the second-generation Range Rover, the original model became known as the Range Rover Classic.

The 3.9-litre engine was in fact not completely new in October 1989, as it had already been standard on US-model Range Rovers for a year. In theory, no doubt the 3.9-litre engine could have been introduced for all markets in 1988, but there were persuasive reasons why it was not. Perhaps the most important of these was that Land Rover wanted to introduce the larger engine at the same time as Discovery in Europe, in order to reinforce the differences between the two models. Also important, however, was the fact that US customers had begun to pile on the pressure for better road performance from their Range Rovers. Land Rover judged that they should not be made to wait until 1989 to get what they wanted (particularly as the Discovery would not be sold in the USA), and so the engine was introduced in two stages.

There was nothing startlingly new about the 3.9-litre engine. The first engines of this capacity had been produced in the late 1970s by John Eales Motors in Coventry for competition use, and before long there were also 3.9-litre engines in Rover SD1 saloons run by the factory's experimental department. By 1986, Land Rover were also looking at these larger engines, and at least one of the Range Rover Olympic prototypes had an experimental example. So there was considerable experience of the 3.9-litre engine by the time it was needed for the production Range Rover.

The production 3.9-litre engine was of course engineered from the beginning to function with the exhaust catalyst already demanded by the US market and expected to become mandatory in Europe during the 1990s. It was simply a large-bore edition of the trusty 3.5-litre engine, but of course was equipped with the latest Lucas 'hot-wire' fuel injection and engine management systems, which metered fuel carefully enough to make average consumption lower than that of the smaller engine.

Although maximum power was increased – to 178bhp in catalysed form and 188bhp in non-catalyst European form – the main benefit of the engine was improved acceleration from higher torque values. The European version had 235lb/ft at 2,600rpm as against the 206lb/ft at 3,200rpm offered by the injected 3.5-litre engine, and the difference showed in the new model's much more vigorous acceleration. However, the changes were not all to the good: the 3.9-litre engine could sound rather strained at high speeds and was in some ways less refined than the injected 3.5-litre type.

In Britain, the arrival of the 3.9-litre engine rather overshadowed the new diesel engine which was announced at the same time. In continental Europe, however, where taxes made the 3.9-litre V8 an expensive irrelevance, the news of a more powerful diesel engine was very well received. The new engine was a derivative of the old, this time with a longer stroke to increase capacity from 2,393cc to 2,498cc – though Land Rover always insisted the capacity was a round 2,500cc. It also had different cylinder heads, improved belt drives and cooling system, modified timing gears and a new timing cover to reduce noise. Land Rover were of course not the only customers for what VM called their HR494 engine (it was already available in the Alfa Romeo 164, for example), but as usual VM had worked with Solihull's engineers to develop a version which suited the Range Rover's characteristics.

The new engine developed 119bhp – just 7bhp more than the older 2.4-litre type – but once again its main benefit lay in improved torque delivery. There was not only more torque than before – 209lb/ft as opposed to 183lb/ft – but it was also developed lower down, at 1,950rpm instead of 2,300rpm. This made the engine much more flexible and much more muscular. Standing-start acceleration times were not a lot better than before, but there was a very real difference in the engine's driveability. As before, the Range Rover Turbo D was available only with the five-speed manual transmission, and initially it was offered only in base-model specification. However, things would soon change: just over a year later, Land Rover brought to market a Vogue Turbo D model with all the refinements of the petrol-powered Vogue except for automatic transmission.

The re-engined 1990-season models also brought with them a revolutionary new anti-lock braking system, which attracted more column inches in the press than either of the two new engines. Available only on the most expensive Range Rover models in each market, this had been developed primarily to keep Range Rover specification levels abreast of those commonly found in the conventional luxury cars which Land Rover now saw as its real competitors.

The Range Rover's ABS had been developed jointly with Wabco, a German company best known for its aircraft and heavy commercial vehicle braking systems. Specifically designed for a four-wheel-drive vehicle which would be used off the road, it was controlled by sophisticated electronics which monitored each wheel independently 250 times a second and could adjust braking pressure up to ten times a second to prevent wheel lock. On the road, it was as

Range Rover 1989–1995

Layout and chassis

Two-door (to 1994) or four-door estate with horizontally split tailgate
Box-section ladder-frame chassis

Engine (petrol)

Type	Light-alloy 90-degree petrol V8
Capacity	3,947cc (4,278cc on long-wheelbase models)
Bore and stroke	93.98mm x 71.1mm (93.98mm x 77mm for 4.2-litre type)
Fuel system	Lucas L-Jetronic fuel injection
Compression ratio	9.35:1 (3.9-litre) 8.13:1 (3.9-litre for USA, 1989–92) 8.94:1 (4.2-litre)
Max. power	188bhp @ 4,750rpm (3.9-litre, 1989–92, except for USA) 178bhp @ 4,750rpm (3.9-litre for USA, 1989–92) 181bhp @ 4,750rpm (3.9-litre, 1992–5) 200bhp @ 4,850rpm (4.2-litre)
Max. torque	235lb/ft @ 2,600rpm (3.9-litre, 1989-92, except for USA) 227lb/ft @ 3,500rpm (3.9-litre for USA, 1989–92) 231lb/ft @ 3,100rpm (3.9-litre, 1992–5) 250lb/ft @ 3,250rpm (4.2-litre)

Engine (VM turbodiesel)

Type	VM type HR 494 OHV in-line four-cylinder turbocharged and intercooled indirect-injection diesel
Capacity	2,500cc
Bore and stroke	92mm x 94mm
Compression ratio	22.5:1
Max. power	119bhp @ 4,200rpm
Max. torque	209lb/ft @ 1,950rpm

Engine (Tdi turbodiesel)

Type	200 Tdi (1992–4) or 300 Tdi (1994-5) OHV in-line four-cylinder turbocharged and intercooled direct-injection diesel
Capacity	2,495cc
Bore and stroke	90.47mm x 97mm
Compression ratio	19.5:1
Max. power	111bhp @ 4,000rpm
Max. torque	195lb/ft @ 1,800rpm

Transmission

Primary gearbox	Five-speed LT77 manual (1989–91) Five-speed LT77S manual (1991–4) Five-speed R380 manual (1994–5) Optional four-speed ZF 4HP22 automatic (standard with 4.2-litre engine)
Ratios	Five-speed (all types, with all petrol engines): 3.32:1, 2.13:1, 1.39:1, 1.00:1, 0.77:1, reverse 3.42:1 Five-speed (with turbodiesel engines): 3.69:1, 2.13:1, 1.39:1, 1.00:1, 0.77:1, reverse 3.42:1

	Four-speed automatic: 2.47:1. 1.47:1, 1.00:1, 0.72:1, reverse 2.08:1
Transfer gearbox	Two-speed Borg Warner
Ratios	High 1.20:1, Low 3.24:1
Centre differential	With automatic locking by viscous coupling
Final drive	3.54:1

Suspension and steering

Front	Live axle with coil springs and hydraulic telescopic dampers, located by radius arms and Panhard rod (air springs instead of coil springs on top models from 1992)
Rear	Live axle with coil springs and hydraulic telescopic dampers, located by radius arms, support rods and central wishbone assembly incorporating a Boge Hydromat self-energizing ride-levelling strut (air springs instead of coil springs on top models from 1992; no self-levelling strut with air springs)
Steering	Power-assisted recirculating-ball, worm and nut
Ratio	17.5:1
Tyres	205 x 16
Wheels	Pressed steel 16in Rostyle type; various alloy options

Brakes

Type	Disc brakes with dual hydraulic line and servo assistance; four-piston callipers on front wheels and two-piston type on rears; internal expanding drum-type parking brake operating on transfer box output shaft
	Wabco ABS on some models
Size	Front 11.75in diameter; rear 11.42in

Vehicle dimensions (in/mm)

Wheelbase	Standard models 100/2,540
	Vogue LSE/County LWB 108/2,743
Overall length	Standard models 176/4,470
	Vogue LSE/County LWB 183/4,648
Overall width	70/1,780
Overall height	70/1,780
Track	58.5/1,480
Kerb weight	Vogue SE 4,379lb/1,986kg
	Vogue LSE 4,739lb/2,150kg

Performance

Max speed	110mph/177km/h (standard models with 3.9-litre engine and long-wheelbase models)
	95mph/153km/h (with VM turbodiesel engine)
	94mph/151km/h (with 200 Tdi or 300 Tdi turbodiesel engine)
0–60mph	11 sec (standard models with 3.9-litre engine and long-wheelbase models)
	15.8 sec (with VM turbodiesel engine)
	16.8 sec (with 200 Tdi or 300 Tdi turbodiesel engine and manual transmission)

impressive as the latest ABS systems in conventional cars, but its most impressive aspect was its off-road operation: in the slimy mud of a downhill slope, it was possible with an ABS-equipped Range Rover to do the off-roader's unthinkable and apply the brakes – whereupon the vehicle would stop, under full control!

With the ABS came ventilated front disc brakes, and these were standardized on all models. Once again, these were items expected in the specification of a high-performance luxury car and so the Range Rover had to have them. Whether they were strictly necessary, though, is perhaps debatable. The argument that they were necessary to dissipate the extra heat generated by the asbestos-free brake pads Land Rover was introducing as part of its new environmentally-responsible policy sounded suspiciously like marketing-speak ...

Changes to the Range Rover for the 1991 season were held over until December 1990, so that the main focus of publicity at Motor Show time could be on the newly-revamped and renamed Defender (Land Rover utility) range and on the new five-door Discovery. In the meantime, the 250,000th Range Rover had rolled off the production lines in September. However, Land Rover did offer a sneak preview of what was to come, at least in the UK. At Motor Show time they announced a limited-edition two-door Range Rover called

Also new for 1990 were ventilated front disc brakes.

This was the era of special editions. For 1991, there was the Range Rover CSK, a specially-equipped two-door model sold in Britain.

the CSK (after Charles Spencer King, the vehicle's originator), and that came as standard with the anti-roll bars which would be standardized on the top-model Range Rovers in December.

The idea of anti-roll bars on the Range Rover was nothing new. Land Rover themselves had looked at an anti-roll suspension to improve the vehicle's handling during the 1970s, and the early 1980s had seen Harvey Bailey Engineering and Warwick Banks Handling offer bolt-on anti-roll kits. However, the big objection to anti-roll bars had always been that they limited the wheel travel which was one of the key factors in the Range Rover's excellent off-road performance.

The about-face had undoubtedly been prompted by market pressures. Buyers of an expensive luxury vehicle like the Range Rover did not expect to be frightened out of their wits whenever they took a corner a little too fast, and so Land Rover had to do something. With a lot of careful development, the suspension engineers headed by Paul Markwick had come up with a pair of anti-roll bars which limited suspension travel by an acceptable half inch. As stiff springing was no longer needed to control body roll, the coil springs were made softer at the same time to give a more compliant ride. It was a considerable achievement, and Land Rover were sufficiently proud of their anti-roll bars to offer them as a conversion for post-1985 models through Land Rover Parts from November 1991.

Various more minor changes characterized the 1991-model Range Rovers, most of them adding to the array of luxury equipment available. But by this stage, Land Rover were preparing for the imminent

The Range Rover CSK

Total built 200
Special features Two-door body
- Beluga Black paint with white coachline
- White CSK decal on scuttle sides
- Bright finish bumpers
- Round driving lamps above front bumper
- Five-spoke alloy wheels with black rims
- Sorrell leather perforated seat facings and door trims
- Leather-covered steering wheel rim
- American Walnut facia rail, console and door trim inserts
- Numbered plaque on radio panel
- ABS
- Anti-roll bars
- Sports dampers
- T-rated tyres
- Engine ECU recalibrated for additional power

The Range Rover CSK was intended primarily to publicize the new anti-roll suspension and was promoted as a sportier Range Rover. Most examples had the 3.9-litre V8 engine and five-speed manual gearbox. Most elements of the CSK's specification were drawn from the four-door Vogue SE or from US-market models.

The basic vehicle was changing along with the special editions. Shortly after the CSK was announced with anti-roll bars, these were fitted as standard to Vogue and Vogue SE models. This shows the bar in place on the rear axle.

For 1991, there was a new deformable plastic fuel tank, and with it a higher-mounted filler...

... while a glass sunroof replaced the steel type.

introduction of a very important new model – a long-wheelbase Range Rover which would set the stage for the larger and more limousine-like second-generation Range Rover then under development.

Once again, the idea of a long-wheelbase Range Rover was nothing new. As Chapter 8 shows, aftermarket conversion specialists had been building stretched Range Rovers since the mid-1970s. There had certainly been discussions about a long-wheelbase four-door model at Solihull in June 1978, and Land Rover did build some long-wheelbase vehicles for special purposes in the late 1980s (*see* Chapter 8). However, the production long-wheelbase model was to have some very special features. Not only would it be laden with all the latest technology, but it would also have an enlarged V8 petrol engine and would feature a revolutionary new air suspension which had been developed in association with Dunlop. It would become the top model in the Range Rover line-up, with a price tag to match, and would demonstrate Land Rover's continuing quest to make the vehicle compete on equal terms with the leading luxury cars of the day.

Everything was on course until the summer of 1991. A date had been set for the

The limited-edition Brooklands was sold in several European countries. This is an export version: UK models did not have the 'Brooklands' decal on the wing.

press launch at the end of October, and the air suspension was undergoing its final reliability tests. Then disaster struck: its electronic control system failed to meet Land Rover's stringent reliability criteria. There was no alternative but to go back to basics, to redesign the system and to put it through the full test sequence once again. There was simply not time to do this before October, and without the air suspension there could be no long-wheelbase Range Rover. So the launch was cancelled at the end of June. No fewer than 102 Range Rovers (many on the standard 100-inch wheelbase) had already been built with air suspension on the assembly lines, and all of those were dismantled. Some sources say the 100-inch models were refitted with conventional steel springs and then sold in the usual way.

That left Land Rover with precious little which was new for the 1992-model Range Rovers. There would be a new edition of the five-speed gearbox with less obstructive synchromesh on first and second gears, but that was hardly the stuff of which headlines were made. There would be some interior upgrades, too, but again nothing to cause great excitement. Nevertheless, Solihull manfully put together a low-key package of improvements and had it ready for announcement in December 1991. Among the changes was a more rigid radio support panel – originally designed to withstand pressure on the air suspension's control switches which were to have been mounted on it.

Meanwhile, TWR Special Vehicles of Kidlington had seen the opportunity to move into the Range Rover aftermarket, and during 1991 they had designed and developed a body kit and alloy wheels for the Range Rover. They offered these to Land Rover, hoping that the manufacturer would agree to offer them as original equipment. Land Rover rejected TWR's rear roof spoiler and rear pillar cover panels (which TWR later marketed independently), but they did show interest in the front and rear aprons, the sill panels and the wheels. First, they asked TWR to redesign the front apron to carry conventional driving lamps instead of the two large circular lamps originally proposed. Then they displayed the body kit on a Range Rover at the Frankfurt Show in autumn 1991. Explained away as a 'styling exercise', it was clearly an attempt to gauge public reaction.

Reaction must have been good, because the TWR styling kit was fitted to a series of limited-edition four-door Range Rovers which were announced at the Geneva Show in March 1992. The British version did not become available until June, was based on the Vogue automatic, and was known as the Range Rover in Brooklands Green; other European countries also had similar models with minor variations in specification. However, there is little doubt that these models were not intended to test the water for the styling kit; their purpose was to offer an interesting and distinctive vehicle to those who had been patiently waiting for the long-wheelbase Range Rover to arrive. In fact, it would be announced just four months later, in October 1992, but few vehicles would actually be available for sale before the end of the year.

The long-wheelbase model with its air suspension was ready this time, and it was introduced as the Vogue LSE in most

Delayed by a year, the long-wheelbase Range Rover arrived in autumn 1992. The longer rear door made the overall proportions look even better, as this US-model Country LWB shows.

Range Rover in Brooklands Green

Total built	150
Special features	Brooklands Green paint
	Five-spoke alloy wheels finished in grey
	Bodykit featuring body-colour front bumper and apron, rear bumper and apron, and side sills
	Top-specification ICE with six-disc CD autochanger and six speakers

The Range Rover in Brooklands Green was promoted as a sporting edition of the four-door model. Specification levels generally were to contemporary Vogue standard, and all 150 examples had the 3.9-litre petrol engine and automatic transmission. The model was originally to be called the Range Rover Brooklands, but that name was cancelled because of a clash with the new Bentley Brooklands model. Similar limited-edition models, but with slightly different specifications, were offered in Europe.

markets, as the Vogue LSEi in Germany, as the Range Rover Vanden Plas in Japan, and as the County LWB in the USA. Land Rover had every confidence that it would be a success in spite of its inflated price tag, not least because the Vogue SE was already accounting for a very high proportion of Range Rover sales by this stage and it was clear that the market wanted more luxury. And so it proved: by the end of the year, one in every four Range Rovers being built was a long-wheelbase model.

The extra length in the wheelbase amounted to just eight inches, which Land Rover argued was as little as was needed to provide limousine-like legroom for the rear seat passengers. Outside the factory, of course, no one knew at this stage that the new 108-inch wheelbase was the same as had been adopted as standard for the second-generation Range Rover then under development ... All the extra length went into the rear door area, so that the number of new panels necessary was minimized. Some commentators argued that the longer rear door made the LSE look even better-balanced than the already handsome 100-inch Range Rover, and there was no doubt that the change had been carried out with some skill.

Extra length brings extra weight and, although the LSE was surprisingly light, Land Rover had taken no chances with the performance of their new flagship model. The LSE therefore came with a 200bhp long-stroke 4.2-litre engine (although the 3.9-litre type was specified for Australia), which produced a maximum speed just higher than that of a 3.9-litre Vogue SE. Even more important was a flatter torque curve, which gave the LSE noticeably improved driveability and made it the fastest accelerating Range Rover to come from Solihull. Yet once again, there was nothing really new under the sun: John Eales Motors had been building 4.2-litre editions of the Rover V8 for many years, and Land Rover did admit that they had

The longer wheelbase gave real lounging room for rear-seat passengers.

Under the bonnet of the long-wheelbase models was a 200bhp 4.2-litre V8, which was not available in other production models. However, Land Rover could be persuaded to fit it to standard-wheelbase models custom-built under the Autobiography scheme after October 1993.

consulted Eales during the development of their own 4.2-litre engine ...

However, the real talking-point on the LSE was undoubtedly the air suspension, which was also standardized on Vogue SE models. Known to Land Rover as ECAS (Electronically Controlled Air Suspension), it was a remarkably sophisticated system. It added refinement by substituting air-filled rubber bags for the steel road springs which transmitted noise from the road wheels to the interior of the car. One penalty was that the airbags did not control body roll as well as conventional springs, even with the anti-roll bars which came as standard, but few buyers noticed. Most of them were more concerned with the system's other advantages.

The two most obvious of these were the most visible: when the vehicle exceeded 50mph (80km/hr), the ECAS lowered the body on its air springs to reduce drag and improve stability, while when the vehicle was at rest, the ECAS lowered the body still further to aid access and loading. Yet these were just two of the five height settings available. The third was normal running height – equivalent to the standard ride height with conventional springs – and above that came a manually selectable 'high' setting, which increased the ground clearance for off-roading.

Right at the top was Extended Profile, which was not a running height at all but rather a feature which automatically

The Electronically Controlled Air Suspension substituted rubber air bags for the steel coil springs of the original Range Rover suspension.

The air suspension brought with it an array of extra switches, seen here on the right of the radio panel in a UK-market Vogue LSE. The switch to the left of the radio and its counterpart on the right control the front seat heaters.

extended the wheels downwards to regain traction if the vehicle had bellied-out while off-roading and the wheels had lost contact with the ground altogether. With the long wheelbase of the LSE, it was necessary if the vehicle was to go anywhere a standard-wheelbase model could; but few owners had the slightest interest in finding that out! Finally, a comprehensive set of computer-controlled overrides and fail-safes, plus a manual override facility, ensured that the ECAS was foolproof.

Over the previous few years, the Range Rover had become increasingly dependent on sophisticated electronics: they took care of engine management, remembered the seat and mirror settings, operated the speedometer, pulsed the brakes when necessary and now adjusted the ride height. So it was no surprise to find further electronic wizardry standardized on the new LSE and the Vogue SE models for 1993. Land Rover employed yet another acronym for this: ETC, which stood for Electronic Traction Control. ETC was actually linked to the ABS system and operated on the rear wheels only. What it did was to pulse the brake on a spinning rear wheel, thus transferring torque to the opposite wheel through the action of the differential. Its primary function was to aid traction on poor surfaces, but it could also prevent wheelspin on metalled surfaces.

With all these new features came a new price tag. The Vogue SE with air suspension was more expensive than the previous year's model, and the Vogue LSE in Britain cost some £3,000 more than that, with a price tag of £38,393 before extras. At this price, the motoring press believed the LSE should offer the first-rate build quality of the luxury saloons from BMW, Lexus and Mercedes-Benz which sold in the same price bracket. Sadly, there was little doubt in their minds that it did not: panel-gaps were both large and uneven, and the dashboard was by this stage looking more and more like the mish-mash it really was, with ill-fitting sections and poor colour matches. These problems were echoed on the standard-wheelbase models, and over the next couple of years the press increasingly pointed out that the Range Rover was failing to meet customer expectations, despite its acknowledged excellence.

Land Rover had recognized that the announcement of the LSE was likely to attract all the headlines, and so they wisely delayed the announcment of 1993's other new model until the month after the Motor Show, even though examples were already being supplied through the showrooms. This second new model was a replacement for the Turbo D, and it was powered by Land Rover's own highly-acclaimed 200 Tdi intercooled turbodiesel engine. Land Rover called it the Range Rover Tdi.

The 200 Tdi engine had first been seen in the Discovery in 1989, and had been made available in the Defender a year later. It was unusual in being a direct-injection design in a market dominated by indirect-injection diesels, but Land Rover had adopted direct injection because of its advantages for fuel economy. The disadvantage was increased combustion noise, and there was no doubt that the 200 Tdi was a raucous engine. However, the turbocharger and intercooler endowed it with prodigious amounts of low-down torque, and the torque spread was wide enough to give good acceleration in the higher speed ranges, too.

Gradual development since 1989 had quietened the engine somewhat, and by 1992 Land Rover were confident enough to introduce it for the Range Rover. In an economy run from Land's End to John O'Groats held shortly after the vehicle's launch, a Range Rover Tdi achieved 53mpg (5l/100km) at an average of 41mph (66km/h) for the 832-mile (1,340km) journey. While most owners were unlikely to see more than 30mpg (9l/100km) in everyday use, this publicity stunt certainly made its point.

There were further minor specification changes when the Tdi was announced. They included a raised loadspace cover (already previewed on the LSE) which

Land Rover's own 200 Tdi engine replaced the VM turbodiesel in 1992. The deeper side sills visible here had been standard since the autumn of 1989; US-market '1990' models built over the summer of 1989 nevertheless had the older sills, as the picture of the early 1990-model County on page 100.

The loadspace cover was redesigned in 1992 to give enough room for suitcases to stand on their ends. Also visible here is the new oddments net at the side of the load bay.

allowed suitcases to be stood on end instead of laid flat, different wood trim and a more comfortable angle of rake for the rear seat backrest. The TWR bodykit became available through Land Rover Parts, together with the star-pattern alloy wheels which TWR had designed in 1991. However, things then went very quiet on the Range Rover front for some seventeen months. Rumours about the forthcoming second-generation Range Rover were rife, and many commentators assumed that the original vehicle had now reached the end of its development life and that Land Rover would not bother to change it any further. As things were to turn out, they could not have been more wrong.

It was odds-on that the second-generation Range Rover would be scooped by one of the spy photographers working for the motoring press before long, and that was indeed what happened. The earliest pictures of the undisguised vehicle were published in September 1993, some six weeks before the Motor Show opened. Inevitably, there was disappointment that the new model was not being announced at the same time, but Land Rover did manage to create a flicker of additional interest in the current Range Rover by announcing a new custom-building service which they called Autobiography.

Promoted as 'the Ultimate Personalization', Autobiography allowed customers to make their own choice from among the features available on the Range Rover, and to combine these with an unlimited choice of paint and interior colours. Their vehicle was then hand-built (or converted from a line-assembled example) by the Special Vehicles division at Solihull. Autobiography did not have an enormous take-up rate, no doubt to the great relief of Special Vehicles who were fully occupied on other projects, but it did at least allow customers to fit a 4.2-litre engine into a standard-wheelbase Range Rover for the first time. Few customers chose colour schemes quite as striking as the vivid blue which distinguished the Motor Show stand Autobiography vehicle, but no doubt there were some unsung lapses of taste as well as some highly desirable one-off vehicles. Vehicles built under this scheme normally had an 'Autobiography' decal on the tailgate.

Starkly simple in many respects, the original Range Rover still looks good a quarter of a century after it was conceived. This is YVB 151 H, chassis number one.

The first few hundred examples had an 8.5:1 compression version of the carburetted 3.5-litre V8 engine. This is one of the very first, pictured in original condition.

Early dashboards were crude by later standards. This one dates from the mid-1970s and shows the cloth upholstery which had by then become available. The footwell coverings were still in PVC, however.

Two-door models pictured on the assembly lines at Solihull in the mid-1970s.

The first turbodiesel-engined Range Rovers appeared in 1986. This is a 1988-model four-door.

H. M. the Queen and H.R.H. Prince Philip aboard the later of the two State Review Range Rovers at the 1994 D-Day Anniversary Celebrations.

Starting in late 1994, several UK police forces took their new Range Rovers in the latest 'Battenburg' livery, which was designed to be more visible in the dark. This one belonged to the Metropolitan Police and has the steel (Discovery) wheels fitted to most new Police vehicles after about 1992.

Police Range Rovers were a familar sight on Britain's motorways. This is one of the last two-door examples to be taken into service by London's Metropolitan Police in the early 1980s.

The Range Rover was a favourite with wealthy clients in the Middle East, who ordered all kinds of luxury conversions from British coachbuilders. This is an Ashton convertible, built in the early 1980s by Glenfrome of Bristol on a 110in (2,794mm) wheelbase chassis.

Luxurious interiors were an important element in the coachbuilders' transformations of the standard Range Rover. This one, again by Glenfrome, is fairly typical.

Coachbuilders were making long-wheel-base Range Rovers many years before Land Rover announced the Vogue LSE (in the US, County LWB). This one was built in the late 1980s by Vantagefield of London and had a wheelbase of 118in (2,997mm).

Range Rovers have always been highly regarded by the French as endurance racers.

France again … and this time a Range Rover on the 1995 Croisière Blanche rally.

The interior of this early 1990s four-door shows how far the Range Rover had come since the days of PVC seats and upholstery some twenty years earlier.

More wood trim than standard, a special steering wheel and red instruments reminiscent of a Ferrari. This is the interior of an early 1990s Overfinch high-performance conversion.

Range Rovers exported to the Gulf states in the late 1980s and early 1990s had grilles and spoilers painted in the body colour. This one was commandeered for military use in Saudi Arabia during the Gulf crisis of 1990–1991.

There have been many special-edition Range Rovers over the years. This is the 1990 CSK, a two-door model named after the Range Rover's originator, Charles Spencer King.

The Range Rover six-wheel Rapid Intervention Vehicle proved a popular airfield fire tender until supplies became difficult in the early 1990s.

The long-wheelbase models provided lounging room in the rear and helped to prepare the market for the second-generation Range Rover which was announced in 1994. This is a 1993-model County LWB, the US-market edition of what most other markets knew as a Vogue LSE.

Another US-specification Range Rover, this time a 1994 model on the standard wheelbase and wearing TWR alloy wheels.

Final revisions before the arrival of the second-generation Range Rover included an impressive new dashboard with air bags, seen here on its announcement at the Geneva Motor Show in March 1994.

The 4.2-litre V8 could *be had in standard-wheelbase Range Rovers built under the Autobiography scheme and these vehicles also had special badging.*

It actually came as something of a surprise when Land Rover announced further changes to the Range Rover in March 1994. In fact, they had for some time been developing a package of changes for both the Discovery and the Range Rover, and announced them both at the same time. Most important of all were the improvements in passive safety made to each vehicle. Both had energy-absorbing crush cans at the front of the chassis rails, and on the Range Rover these were concealed by small protrusions like over-riders on the bumper. Both also had side impact beams in their front doors, and the LSE also had such beams in its rear doors. Lastly, both could be equipped with airbags, which demanded a completely redesigned dashboard.

The new Range Rover dashboard was similar to the new one in the Discovery, but there were differences. It consisted of a single-piece moulding –at last eliminating the patchwork-quilt effect of the old facia – with vastly improved rotary heater controls which fronted a new heater box. Other interior trim had been changed to match, and the multi-function column stalks were replaced by equivalent items borrowed from

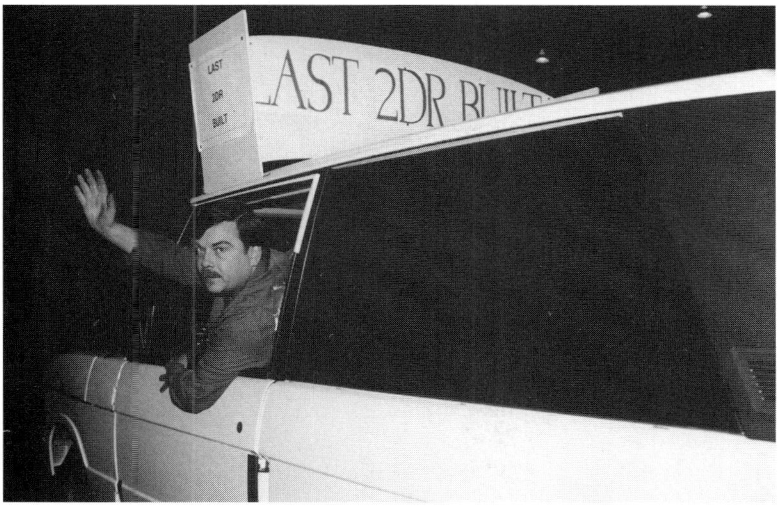

The last two-door Range Rover was built on 11 January 1994 and was a van variant destined for Portugal (see *Chapter 6).*

The 1995-model Range Rovers actually appeared in March 1994. This one was pictured at the Geneva Motor Show and clearly shows the new 'crush cans' on the front bumper. Just visible are the new black-shaded edges to the windscreen.

The most striking new feature in March 1994 was the completely redesigned dashboard, incorporating air bags. Here it is in a Vogue SE. Note the additional speakers mounted on the windscreen pillars.

the Rover 800 saloon. Unfortunately, the new facia was not an unqualified success. The new steering wheel with its large boss containing the airbag tended to obscure some of the instruments, and the interior felt more claustrophobic, partly because of dark shading around the windscreen edges which had been introduced primarily to give the screen a less angular appearance from outside.

With these changes came a modified version of the turbodiesel engine, now known as a 300 Tdi and redeveloped to give quieter running and meet forthcoming new diesel emissions legislation in Europe. The LT77S gearbox was replaced by a stronger gearbox called the R380, again five-speed but with a very much slicker change action. Other changes affected the model mix: automatic transmission became available with the Tdi engine for the first time (there had been an automatic Tdi Discovery since September 1993), while it was now possible to order a Vogue SE with diesel power.

These, then, were the models which remained available when the second-generation Range Rover was announced at the end of September 1994. A restricted model range was offered, consisting in the UK of just five models: Vogue Tdi manual and automatic, Vogue Tdi SE automatic, Vogue V8 automatic and Vogue SE V8 automatic. All were now known as Range Rover Classic models, to distinguish them from the second-generation Range Rovers, and all wore a small 'Classic' badge beneath the Land Rover oval on their tailgates.

Land Rover's intention was to keep them in production as long as demand continued – and in fact for a time the Classic was still

The March 1994 changes also brought the improved 300Tdi version of Land Rover's turbodiesel engine.

When the second-generation Range Rover was launched in autumn 1994, the old model was rebadged as Range Rover Classic.

the only Range Rover on sale in some areas, such as the USA, where the new model was not introduced until January 1995. And so in this guise, the Range Rover celebrated its twenty-fifth full year of production, already superseded but still popular enough for its makers to justify continuing production on a reduced scale.

This 1994-model Vogue SE has the optional TWR-designed body styling kit and TWR alloy wheels.

During 1995 Range Rover Classic production was moved to a dedicated single assembly line in a stand-alone facility in the old South Block. Production had dwindled to sixty-nine vehicles a week by August, and was expected to end in February 1996.

The US Range Rover, 1989–1995

The US market increased in importance during this period, in 1990 becoming the largest of the 80 overseas markets to which the Range Rover was exported.

Brief specification details of the 1989–1995 US model Range Rovers are as follows. Some US dealers also produced their own special-edition models during this period.

October 1989 to October 1990

ABS braking system with ventilated front discs standard; high-mounted stoplight on tailgate. All standard models now had leather upholstery in tan only; no fabric option.

County model introduced with special badging on tailgate, radio surround panel and wheel centres, and 'County' embossed on front seat head restraints. County had five new micatallic paint colours, plus Beluga Black. Grey interior trim standard with certain colours; fabric upholstery optional. Lateral stitching pattern for leather. Burr walnut trim on facia, doors and console. Chrome bumpers and special metallic wheel finish. Option pack of electric sunroof and self-dipping rear view mirror. High-grade ICE system with upgraded speakers and six-disc CD changer.

In the USA, the 1990 season saw the arrival of the Range Rover County. The bright bumpers, body-colour wheels and special wheel centres are visible in this picture.

GDE (Great Divide Edition) introduced in July 1990; limited edition of 400 to commemorate Range Rover expedition along the US Continental Divide in August 1989; 410 actually made. Alpine White paint with beige perforated

The 1990 model Range Rovers were introduced to US journalists on the Great Divide Expedition, an epic off-road safari along the US Continental Divide. For 1991, a commemorative model called the Great Divide Edition or GDE was put on sale.
It featured a bull bar as standard, no front spoiler, five-spoke wheels (earlier seen as accessories for the Discovery) and a special interior.

The GDE featured special tailgate badging.

leather upholstery and polished American Walnut wood trim. Sunroof and self-dipping rear view mirror, smoke grey five-spoke alloy wheels, front bumper/brush bar and rear light guards all standard. No spoiler. Special tailgate badge and numbered brass plate on driver's door edge.

November 1990 to October 1991
New three-model range: standard Range Rover supplemented by Hunter and County SE. Anti-roll bars on standard and County SE models; larger fuel tank with resited filler and remote filler door release. Heated door locks, improved sound proofing on bulkhead and centre console, self-dipping rear view mirror now standard, tool storage bag in spare wheel well. Black bumpers on all models; metallic gunmetal grey wheel finish on standard models. A 120-watt, seven-speaker ICE system standard; CD autochanger optional. Three new paint finishes. Optional electric glass sunroof with sunblind.

Hunter specification included dog guard, luggage net, silver-painted alloy wheels and special badging. Hunter lacked spoiler, ABS, anti-roll bars, loadspace cover and power-adjustable front seats, and was available only in four solid paint colours. There were 120 Hunters sold in 1990 and 405 sold in 1991, making a total of 525.

County SE finished in Beluga Black with tan seats, silver five-spoke alloy wheels, aluminium brush bar with removable headlamp guards and electric sunroof all standard. County SE badges fitted.

November 1991 to September 1992
Two-model range, consisting of standard Range Rover and County. County added ABS, anti-roll bars and CD autochanger to specification of standard model. Beluga Black paint an extra-cost option on County, but not available on standard model.

October 1992 to September 1993
Two-model range, consisting of County and County LWB. County LWB specification as for European Vogue LSE except for bright metal bumpers. County specification included ABS, ETC (electronic traction control), and anti-roll bars as standard; 3.9-litre engine now had same

specification as European versions. County available in four colours only.

October 1993 to December 1994
Two-model range, consisting of County and County LWB. County now had air suspension and bright-finish TWR five-spoke alloy wheels as standard. County LWB had 'Cyclone' alloy wheels with Quicksilver finish as standard. Revised paint and trim colour options.

County models no longer available after March 1994 approximately, leaving only County LWB on sale.

Limited edition of 200 25th Anniversary models announced in July 1994, based on County LWB; 250 actually made. Aspen Silver paint and Ash Grey leather trim with Dark Granite piping; TWR five-spoke wheels with sparkle-silver finish. Embossed Anniversary logo on front head restraints and '25th Anniversary Edition' jewel badge on tailgate.

January 1995 on
Single-model range (alongside second-generation Range Rover 4.0SE). Standard-wheelbase model reintroduced as Range Rover County Classic; specification generally similar to European Range Rover Classic in Vogue SE form but with bright metal bumpers, TWR five-spoke wheels and 160-watt nine-speaker ICE system with optional CD autochanger.

European Special Editions

Range Rover marketing became more adventurous in Europe after the 1989 arrival of Discovery, and several special-edition models were created. In all cases, these had special combinations of optional features, sometimes enhanced by non-standard features added by the importers. This list is representative rather than definitive.

Ascot Germany, 1992. Based on four-door Vogue SEi, equipped with TWR bodykit (as UK-market Range Rover in Brooklands Green), and finished in Beluga Black.

Balmoral France and Spain, 1993 model-year. Two-door diesel (also available as a four-door in France) with air conditioning, six-disc CD autochanger, dual coachline and 'Balmoral' crest decals on scuttle sides.

Brooklands Holland and Switzerland, 1992. Essentially the same as the UK-market Range Rover in Brooklands Green. Sixty-five examples were sold in Switzerland.

Countryman Spain, 1993 model-year. Two-door diesel with bull bar and light protectors, Hella long-range driving lamps, dog guard, rear parcel shelf, waterproof seat covers, waterproof load-space liner, side-steps, single coachline and 'Countryman' decal badges on scuttle sides.

Silver Fox France, 1992 model-year. Two-door in Aspen Silver with coachline and 'Silver Fox' decals on scuttle sides.

Silver Spring France, 1992 model-year. Four-door petrol without spoiler. Double coachline and 'Silver Spring' decals on scuttle sides.

Vogue S Italy, 1993 model-year (from December 1992). Four-door diesel with ABS, air conditioning, leather upholstery, electric sunroof, roof bars, shaded five-spoke alloy wheels, additional driving lamps in grille, high quality audio system, and TV for rear seats.

20th Anniversary Switzerland, 1990 model-year (from July 1990). Four-door automatic, finished in Westminster Grey or Roman Bronze, with bright finish bumpers, SE-style coachline and '20th Anniversary' logo. Additional wood trim on dashboard and centre console. 60 examples only.

3.9Ei Portugal, 1993 model-year (June to October 1993). Two-door petrol, equipped with TWR bodykit and alloy wheels. Sixty examples only, equipped by Rover Portugal.

Key Production Changes, 1989–1995

NB Paint and trim colour option changes are listed separately.

October 1989	3.9-litre fuel-injected V8 replaced 3.5-litre type
	2.5-litre VM turbodiesel replaced 2.4-litre type
	New gearwheels in manual gearboxes, with wider faces and forged dog-teeth
	Manual gearboxes fitted with oil pump and oil cooler (petrol models only)
	Automatic gearboxes fitted with top-to-third kickdown inhibitor
	Ventilated front brake discs replaced solid type; asbestos-free pads standardized
	Multi-point central locking replaced early type operable only from driver's door
	Front spoiler now had six cooling slots in bottom edge instead of four (Vogue SE only)
	Land Rover oval logo added to grille
	Vogue SE decals added to tailgate where relevant; vehicles badged as Vogue SEi in some European countries
	Improved seat padding
	Lateral-stitch upholstery pattern
	Four-vent central rail (formerly only used with air conditioning) now standardized
	Revised instrument faces and warning light cluster
	Electronic speedometer
	Transfer box temperature warning light added to facia
	Vogue models now fitted with former Vogue SE levels of soundproofing
	New options: ABS (standard on Vogue SE)
	Exhaust catalyst (V8-engined models only)
	Heated windscreen
September 1990	CSK limited edition announced
December 1990	Vogue Turbo D model announced
	Front and rear anti-roll bars standardized (Vogue and Vogue SE models only)
	Enlarged (18-gallon/80-litre) polyurethane fuel tank
	Fuel filler neck and flap relocated higher on rear wing
	Six-slot spoiler standardized on Vogue and Vogue Turbo D models
	Heated front door locks standardized (four-door models only)
	Glass sunroof replaced metal type (Vogue SE models only)
	Sound insulation improved
	Recessed seat belt anchorage points in rear floor replaced protruding type
	New ICE speakers with improved bass response
	Electrically operated automatic anti-dazzle mirror standardized (Vogue SE models only)
	Cruise control standardized (Vogue SE automatic models only)
	New option: Five-spoke alloy wheels
December 1991	LT77S five-speed gearbox replaced LT77 type
	Restyled heater surround panel
	Reading lights incorporated in anti-dazzle rear view mirror

	(Vogue and Vogue SE models only)
	Land Rover oval badge added to tailgate
	New style of decal badging for Vogue SE, with upper and lower case letters instead of all upper case
	New options: Heated front seats with two-position memory for seat and door mirrors (standard on Vogue SE)
	Leather upholstery available with non-electric seats (Vogue models only) Clarion radio/cassette with six-disc
	CD autochanger and six speakers (Vogue and Vogue SE models only)
June 1992	Range Rover in Brooklands Green limited edition model announced
October 1992	Vogue LSE model announced
	Raised letters for bonnet and tailgate badges on all models
	Exhaust catalyst standardized with V8 petrol engines
	Electronic Traction Control standardized with ABS (Vogue, Vogue SE and Vogue LSE only)
	Electronically Controlled Air Suspension standardized on Vogue SE and Vogue LSE
November 1992	200 Tdi intercooled turbodiesel engine replaced 2.5-litre VM turbodiesel; diesel model renamed Range Rover Tdi
	Raised loadspace cover standardized
	Greater rake for rear seat backs
	Italian Poplar wood trim replaced American Walnut
	New options: TWR star-pattern alloy wheels
	Body styling kit
January 1994	Two-door bodyshell and Range Rover Van ceased production
March 1994	Vogue SE turbodiesel model announced
	300 Tdi intercooled turbodiesel engine replaced 200 Tdi in Range Rover Tdi
	R380 five-speed gearbox replaced LT77S type
	Cyclone alloy wheels standardized on Vogue LSE
	Bodykit standardized on Vogue LSE
	Crush cans on front bumper
	Side intrusion bars fitted inside front doors of all models and rear doors of Vogue LSE
	Completely restyled dashboard with driver and passenger airbags
	Improved heater with new rotary controls
	Vertical pleated upholstery replaced horizontal pleated type
	New option: Automatic transmission made available with Tdi engine
October 1994	Range Rover renamed Range Rover Classic
	'Classic' badge added to tailgate
	Restricted model-range available (exact details depending on country)

Paint and Trim Options, 1989–1995

October 1989 to November 1990

There were thirteen standard paint colours, of which six were micatallics (indicated by 'mca'), one was metallic (indicated by 'm'), and one was a clear-over-base type (indicated by 'cob'). Vogue SE models were available in only six colours (Ardennes Green, Beluga, Cairngorm Brown, Chamonix White, Plymouth Blue and Westminster Grey). One colour (Westinster Grey) was exclusive to the Vogue SE while one other (Beluga) was available only on the Vogue SE and the limited-edition CSK. Alloy wheels were now finished in Gunmetal. Standard upholstery was in Brogue or Winchester cloth; Vogue SE models came with Winchester or Sorrell leather upholstery; the CSK had perforated Sorrell leather upholstery. The options were:

Ardennes Green (mca)	with trim in	Brogue or Winchester
Arles Blue		Winchester
Arran Beige		Brogue
Beluga (cob)		Sorrell or Winchester
Cairngorm Brown (mca)		Brogue or Sorrell (Vogue SE)
Chamonix White		Brogue or Winchester; Sorrell or Winchester (Vogue SE)
Clearwater Blue (mca)		Winchester
Colorado Silver (m)		Brogue or Winchester
Eastnor Green		Brogue
Plymouth Blue (mca)		Winchester
Portofino Red		Brogue or Winchester
Trocadero Red (mca)		Brogue or Winchester
Westminster Grey (mca)		Winchester

December 1990 to October 1991

There were sixteen standard paint colours, of which seven were micatallics, two were metallics and one was a clear-over-base type. Vogue SE models were available in only six colours (Alpine White, Ardennes Green, Aspen Silver, Beluga Black, Plymouth Blue and Westminster Grey). Three colours (Aspen Silver, Beluga Black and Westminster Grey) were exclusive to the Vogue SE. Standard upholstery was in Brogue or Winchester cloth; Vogue SE models came with Winchester or Sorrell leather upholstery. The options were:

Alpine White	with trim in	Brogue or Winchester
Ardennes Green (mca)		Brogue or Winchester; Sorrell or Winchester (Vogue SE)
Arles Blue		Winchester
Arran Beige		Brogue
Aspen Silver (mca)		Sorrell or Winchester
Beluga Black (cob)		Sorrell or Winchester
Cairngorm Brown (mca)		Brogue
Clearwater Blue (mca)		Winchester
Colorado Silver (m)		Winchester
Eastnor Green		Brogue
Pembroke Grey		Winchester
Plymouth Blue (mca)		Winchester
Portofino Red		Brogue or Winchester
Roman Bronze (m)		Brogue

Trocadero Red (mca)		Brogue or Winchester
Westminster Grey (mca)		Winchester

November 1991 to September 1992
There were sixteen standard paint colours, of which seven were micatallics, two were metallics and one was a clear-over-base type. Vogue SE models were available in all listed colours, but body-colour wheels were available only with Aegean Blue, Ardennes Green, Aspen Silver, Beluga Black, Plymouth Blue and Westminster Grey; with other colours, the wheels were Gunmetal Grey. Three colours (Alpine White, Beluga Black and Westminster Grey) were exclusive to the Vogue SE, and of these Alpine White could be had only in the Middle East. One colour (Brooklands Green) was exclusive to the Brooklands Green limited edition. Standard upholstery was in Brogue or Winchester cloth; Vogue SE models came with Winchester or Sorrell leather upholstery. The options were:

Aegean Blue (mca)	with trim in	Winchester
Alpine White		Sorrell or Winchester
Ardennes Green (mca)		Brogue, Sorrell or Winchester
Arles Blue		Winchester
Arran Beige		Brogue or Sorrell
Aspen Silver (m)		Brogue, Sorrell or Winchester
Beluga Black (cob)		Sorrell or Winchester
Brooklands Green		Winchester
Cairngorm Brown (mca)		Brogue or Sorrell
Clearwater Blue (mca)		Winchester
Eastnor Green		Brogue or Sorrell
Pembroke Grey		Winchester
Plymouth Blue (mca)		Winchester
Portofino Red		Brogue, Sorrell or Winchester
Roman Bronze (m)		Brogue or Sorrell
Trocadero Red (mca)		Brogue, Sorrell or Winchester
Westminster Grey (mca)		Winchester

September 1992 to September 1993
There were fifteen standard colours available, of which seven were micatallics, two were metallics and one was a clear-over-base type. Vogue SE models were available in six colours (Alpine White, Ardennes Green, Aspen Silver, Beluga Black, Plymouth Blue and Westminster Grey), all with body-colour wheels. Vogue LSE models were available in six colours (Alpine White, Ardennes Green, Beluga Black, Mosswood, Plymouth Blue and Roman Bronze), all with body-colour wheels. Beluga Black was available only on Vogue SE and LSE models; Mosswood was exclusive to Vogue LSE models; and Westminster Grey was exclusive to Vogue SE models.

Standard upholstery was in cloth; leather was standard on Vogue SE and LSE models and optional on Vogue models. Upholstery colours were grey (Winchester in cloth and for Vogue and Vogue SE leather; Dark Sable for Vogue LSE leather) or brown (Brogue for cloth, Sorrell for Vogue and Vogue SE leather, Saddle for Vogue LSE leather). The options were:

Aegean Blue (mca)	with trim in	grey
Alpine White		brown or grey
Ardennes Green (mca)		brown or grey
Arles Blue		grey
Aspen Silver (m)		brown or grey
Beluga Black (cob)		brown or grey

Eastnor Green	brown
Mosswood (mca)	brown
Pembroke Grey	grey
Plymouth Blue (mca)	grey
Portofino Red	brown or grey
Roman Bronze (m)	brown
Sonoran Brown (mca)	brown
Trocadero Red (mca)	brown or grey
Westminster Grey (mca)	grey

October 1993 to March 1994
The standard paint and trim colours remained unchanged. However, Land Rover's new Autobiography service enabled customers to order vehicle with paint and interior trim entirely to choice.

March 1994 to May 1995
There were fourteen standard paint colours, of which six were micatallics, three were metallic, and one a clear-over-base type. Vogue SE models were available in seven colours (Alpine White, Ardennes Green, Aspen Silver, Avalon, Beluga Black, Niagara and Plymouth Blue), all with body-coloured wheels. Vogue LSE models were available in eight colours (Alpine White, Ardennes Green, Avalon, Beluga Black, Montpellier, Mosswood, Niagara and Plymouth Blue), all with sparkle-finish Cyclone alloy wheels. Two colours (Beluga Black and Niagara) were available on Vogue SE and LSE models only; one colour (Mosswood) was exclusive to Vogue LSE models.

Standard upholstery was in velour; leather was standard on Vogue SE and LSE models and optional on Vogue models. Upholstery colours were light grey (Granite in velour or leather), dark grey (Ash Grey in leather only) or brown (Brogue in velour or Sorrell in leather).

Under the Autobiography scheme, it was possible to order a vehicle in any colour and with any trim colour. However, the standard options were:

Aegean Blue (mca)	with trim in	brown, dark grey or light grey
Alpine White		brown, dark grey or light grey
Aspen Silver (m)		dark grey or light grey
Ardennes Green (mca)		brown, dark grey or light grey
Arles Blue		brown or light grey
Avalon (mca)		dark grey or light grey
Beluga Black (cob)		brown, dark grey or light grey
Coniston Green		brown
Montpellier (mca)		brown, dark grey or light grey
Mosswood (mca)		brown, dark grey or light grey
Niagara (m)		dark grey or light grey
Plymouth Blue (mca)		brown, dark grey or light grey
Portofino Red		brown or light grey
Roman Bronze		brown or light grey

Chassis Numbers and Production Figures, 1989–1995

Range Rover chassis numbers continued to follow the seventeen-digit VIN system in this period, with the final six digits being the actual serial number. After December 1990, Range Rovers no longer had their own serial number sequence but shared a common sequence with Defender and Discovery models.

A typical number would be SALLHAMN7GA460357. This decodes as follows:

1. SA	World manufacturer code for Rover Group			
2. L	Manufacturing sub-division (Land Rover Ltd)			
3. LH	Model code (Range Rover)			
4. A	First type code	A	=	100-inch wheelbase
		B	=	108-inch wheelbase
5. M	Second type code	B	=	two-door
		M	=	four-door
6. N	Engine type	F	=	2.5-litre Tdi turbodiesel
		M	=	3.9-litre injected V8
		N	=	2.5-litre VM turbodiesel
		3	=	4.2-litre injected V8
7. 7	Steering and transmission code	3	=	RHD, automatic
		4	=	LHD, automatic
		7	=	RHD, 5-speed manual
		8	=	LHD, 5-speed manual
8. G	Model-year code	G	=	1990 models
		H	=	1991 models
		J	=	1992 models
		K	=	1993 models
		L	=	1994 models
		M	=	1995 models
9. A	Assembly location	A	=	Solihull
		F	=	overseas (shipped as CKD)
10. 460357	Serial number			

Note US-market vehicles for 1992 and later model-years have both EEC VIN codes as above and a second VIN code to meet US regulations. The serial numbers are the same.

Land Rover Ltd quote the following production figures for Range Rovers in this period. These figures equate to calendar year rather than to model-year.

1989	28,513
1990	24,202
1991	16,408
1992	15,334
1993	14,870
1994	15,283

6 Conversions: Commercial, Military and Emergency Service Range Rovers

It has always been part of the Land Rover tradition that the basic vehicle should be adapted to suit as many different rôles as possible. Despite doubts in some departments at Solihull about the Range Rover's right to be considered a Land Rover at all, it was not long before a policy decision was made to examine the scope for adaptations and conversions. In fact, there were already plans to turn it into a Police patrol car by the end of 1968, some six months before a definitive prototype had been built and more than 18 months before the model's public launch!

Conversions for the emergency services were already an important part of the Land Rover market, and it was perhaps only natural that Solihull should start to think of adapting the Range Rover to suit that market as well. In fact, the Range Rover would offer at least one very useful advantage in that its road performance was very much better than that of existing Land Rovers; as a fast-response vehicle, it therefore looked like being ideal for the emergency services. As a result, George Mackie's Land Rover Special Projects Department started to look at the vehicle's possibilities just as soon as some pilot-build examples became available in the early months of 1970.

However, it was not Special Projects' policy back in 1970 to carry out the conversions themselves. Their function was to advise, approve and co-ordinate the work of conversion specialists outside the company, and their initial efforts with the Range Rover were mainly feasibility studies. Once a way forward had been identified, the actual conversion work was farmed out to a number of recognized specialists to whom the coveted Land Rover Approval was granted. What that meant was that Special Projects had evaluated the conversion and were satisfied that it did not compromise the integrity of the base vehicle, and that Solihull would therefore honour its standard warranty.

During the early 1990s, Solihull restricted the supply of chassis/cab Range Rovers, mainly to protect the upmarket image of the Range Rover; as a result, very few six-wheel conversions were carried out during this period. At least one major converter was obliged to buy complete vehicles through dealerships in order to fulfil orders, and this inevitably put the price of the conversions up by a huge margin. Meanwhile, the emergency-vehicle market was gradually being taken over by the Discovery, which was of course cheaper and therefore appealed to fleet buyers. However, examples of Police Range Rovers continued to appear well into the Range Rover's twenty-fith year of production.

POLICE RANGE ROVERS

The initial interest in the Range Rover as a Police patrol vehicle came from the Warwickshire Constabulary during 1968. That force's Traffic Superintendent approached Solihull about the purchase of some vehicles for high-speed motorway patrol duties, and was gently steered in the direction of the forthcoming Range Rover. At a meeting held on 18 December 1968, the Sales Division were given permission to discuss the vehicle in detail with the Police, and from then on almost everything was plain sailing. Tony Poole of the Styling Department sketched up a possible Police variant in May 1969, and Special Projects had prepared the first Police demonstrator by September 1970.

This first demonstrator had a number of unsatisfactory features. The rear side windows had been replaced by roller shutters which gave access to equipment racks in the rear, but these were found to be too high up for ease of use. In addition, some equipment still had to be stowed on the floor and could only be reached through the open tailgate. So an improved configuration was drawn up, in which the vehicle retained its rear side windows but was fitted with equipment racks on the rear load floor. Access was then through the tailgate, and careful design of the racking made everything easily accessible.

Police forces still dispute which of them was the first to take Range Rovers on their strength, but it seems probable that the honour fell to the Cheshire Police, whose first vehicles were a batch of four, registered as STU 699 J to STU 703 J in April 1971. Just a month later, the Staffordshire Police put into service their first motorway patrol Range Rover, registered YKN 631 J. The Kent and Warwickshire forces also took early examples, although in each case these had the K-suffix registrations first issued in August 1971.

Range Rovers soon became a common

The Police Range Rover became a familiar sight in Britain. This is a mid-1970s example which was operated by the Cleveland Police. On the roof is a Dale Stem-Lite, which extended upwards on a telescopic mast to provide floodlighting for an accident scene at night.

sight in the livery of a Police motorway patrol car and few forces in Britain managed without at least one or two on the strength of their Traffic Division. Sales were undoubtedly helped by a favourable article in *Police Review*, where W. R. Taylor wrote,

> Unsurpassed versatility is perhaps the shortest and most complete summing-up that could be made about the Range Rover. Few vehicles have caused such a stir in Police circles when the details were only rumoured, and now that the full specification is available for all to see, the stir is even greater – and with just cause. In all the years I have been road-testing vehicles I have never come across such a universally-acceptable, or ready-made, Police vehicle as the Range Rover.

The reasons why the Range Rover had such great appeal are not hard to discover. First, it was fast enough to reach the scene of an incident quickly; second, it could be taken off-road to reach an otherwise inaccessible location or simply to get around an obstacle; and third, it had enough load space to carry emergency equipment. The Police soon discovered that it was also strong enough to tow broken-down articulated lorries off the main carriageway of a motorway, although such use did tend to wear the clutch prematurely!

Solihull even had some success in selling the Range Rover to overseas Police forces for similar duties. The Swiss were among the first customers outside Britain, and by the end of the 1970s it had also entered service with the Belgian Gendarmerie, the Cyprus Police, the Dutch Rijkspolitie (State Police), the Liechtenstein Police, the Royal Malaysia Police, the Oslo City Police, the South Australia Police, the Tasmania Police and the Utrecht City Police. The Bahrain Police had also taken some Range Rovers, converted to 110-inch wheelbase by FLM Panelcraft (*see* Chapter 8).

When the Range Rover was upgraded in autumn 1979, Land Rover were concerned

Most Police Range Rovers in Britain were used for motorway patrol work and carried a large quantity of equipment for dealing with traffic incidents. This 1981 example was owned by London's Metropolitan Police and was probably among the last two-doors to enter service. Most Police Forces took four-door models as soon as they became available.

Some overseas Police Forces were not prepared to wait for the factory four-door, however. This example was one of several converted by FLM (Panelcraft) of London for the Bahrain Police in the mid-1970s, and featured a 110-inch (2,794mm) wheelbase as well as the extra doors.

that its increased cost would deter fleet users like the Police. They therefore introduced the cheaper low-specification Range Rover Fleet Line (*see* Chapter 3). However, the Fleet Line was not a great success, and was finally phased out around 1984, by which time Police forces were already switching over to four-door Range Rovers.

Throughout the 1980s, British Police forces continued to take delivery of new Range Rovers, although they naturally took the cheapest versions Solihull could provide. This meant that they had manual windows for as long as these remained available, pressed-steel Rostyle wheels long after these had been replaced by three-spoke alloys on all UK-market showroom models, and a variety of other deletions from the specification. During the 1990s, Police Range Rovers took on the stylish pressed-steel disc wheels of the Discovery (never available on a showroom-specification Range Rover) and a number of forces ordered them with the Tdi turbodiesel engine as soon as that became available.

Although Solihull did develop a standard Police specification for the Range Rover, no two Police forces' vehicles were ever exactly alike. Different operating conditions and the differing opinions of Divisional Superintendents in charge of Traffic ensured that special features were

Basic Police Specification – 1972

* 20 ACR, 60-amp alternator (battery sensed)
* Split charge facility for radio communications
* Electrical harness for roof auxiliaries
* Additional battery with clamps and cables
* Battery condition meter (switchable to either battery)
* Auxiliary switch panel with provision for six three-position switches and two warning lights
* Fusebox for auxiliary switches
* Calibrated speedometer in dash centre; split drive option to retain operation of vehicle speedometer

Basic Police Specification – 1992

* Twin batteries with split charge facility
* Map reading light
* Calibrated speedometer
* Provision for Police radio
* Steel wheels (Discovery type)

This Metropolitan Police vehicle may look like a standard four-door Range Rover, but it is actually an armour-plated vehicle used for special duties. The steel disc wheels give away what a closer look confirms to be the case.

added to the 'standard' specification, often in a force's own workshops rather than by Solihull. In Britain, the Ministry of Defence Police ran some armoured Range Rovers, and the Metropolitan Police had some unmarked armoured examples on both the standard and 110-inch wheelbases (*see* Chapter 8) for VIP protection duties. Some of the Metropolitan Police examples had partial armour only, covering the front of the vehicle and the windscreen.

Many Police Range Rovers were in constant use for nearly 24 hours a day, taking time off duty only for essential servicing, and as a result many of them racked up enormous mileages. Reliability was generally reported as excellent, and some ex-Police Range Rovers eventually reached the civilian market with several hundred thousand miles behind them, ready for a new lease of life. Typically, a Police Range Rover in Britain would be replaced after four or five years' service, but many lasted much longer than that. The Metropolitan Police, for example, were still running at least one 13-year-old example in 1995.

AMBULANCE CONVERSIONS

Over the summer of 1970, the Special Projects Department took a first look at developing a Range Rover ambulance conversion. The biggest difficulty was to find enough space within the body and, to make room for a stretcher to be carried, the front passenger seat had to be removed and then repositioned up against the facia, facing backwards. This was no doubt rather

The 110-inch (2,794mm) ambulance designed by Solihull in conjuction with Spencer Abbott was not a success.

uncomfortable, but there were worse problems. The limited headroom in the vehicle meant that the medical attendant who used that seat had to crawl on his knees to tend the patient, while the high floor level in the back of the vehicle made it difficult to load the stretcher trolley. A first prototype, registered ELA 830 J, was built with this configuration, but soon proved unsatisfactory.

The only answer was to enlarge the vehicle, and so in February–March 1971, Special Projects cut a pilot-production Range Rover in half and stretched its wheelbase by 10 inches. A simple extra panel was inserted into the body to join the front and rear halves, and the vehicle then went out on road-test. Everything proved satisfactory, and so YVB 158 H (chassis number 35500010A) was then passed on to ambulance specialists Wadham Stringer, of Waterlooville in Hampshire, and by September it had been fitted with a special high-roof ambulance body. This was designed to carry two stretchers, and additional interior space had been created by allowing the rear of the body to overhang the chassis by several inches.

The Wadham Stringer design was

Altogether more successful were ambulance bodies on the 110-inch chassis by Herbert Lomas ...

... and Pilcher-Greene, this one for a Middle Eastern customer.

granted Land Rover Approval, but the nature of the ambulance market was such that Solihull could not afford to give one company sole approval. So it was that two further ambulance specialists, Herbert Lomas of Congleton in Cheshire and Pilcher-Greene of Burgess Hill in Sussex, developed their own bodies for the 110-inch wheelbase chassis. Their prototypes, very similar in concept to the Wadham Stringer design, were ready by the middle of 1972 and went on sale shortly afterwards. All three companies continued to build ambulance bodies on the stretched-wheelbase Range Rover chassis throughout the 1970s and 1980s, selling mostly overseas but occasionally to British users. During the 1980s, Herbert Lomas closed down; its products and licences were taken on by MMB International Ltd of Macclesfield, but the new company did not resume production of Range Rover ambulances.

There was one other early proposal for an ambulance body which failed to find a market. This was a dual-purpose design drawn up by the Styling Department at Solihull, a high-roof body for the 110-inch wheelbase which could be fitted out as either a 12-seater Station Wagon or as an ambulance. A prototype body was built by Spencer Abbott, the Birmingham conversion specialists, and was fitted to YVB 168 H (chassis number 35500025A). A left-hand drive ambulance demonstrator was then built to this design and registered as YXC 999 K before being taken to a medical exhibition in Switzerland.

However, the body did not prove to be a success; there may have been one more ambulance version built in 1976, but this was probably later converted to the mobile laboratory which was used at British

However, the contract for military ambulances for the British forces went to Wadham Stringer. Both right-hand drive and left-hand drive types were ordered. The first ones entered service in 1977, but this example dates from about 1983.

Leyland's Gaydon test centre and registered as UDU 924 R. Ironically, it was later sold to a company specializing in rally medical cover and kitted out as an ambulance! The only Station Wagon made was an eight-seater delivered to construction mogul Sir Alfred McAlpine some time around November 1973 and registered as MCA 723. It seems to have been used by shooting parties. Some Rover car brochures show a mobile laboratory registered as SVC 623 X on the Gaydon test track, but this 1982 registration number was almost certainly a fake added to UDU 924 R, as the Spencer Abbott conversion was no longer listed after 1978. Nevertheless, Spencer Abbott did do well out of their involvement with the project: they were awarded the contract to carry out all the Range Rover 10-inch chassis extensions on Solihull's behalf.

Even the 110-inch wheelbase was not enough for some customers. During 1975, a Norwegian conversion specialist stretched a standard 100-inch wheelbase Range Rover chassis by a full metre (39.4 inches)

Solihull examined this Norwegian-built long-wheelbase ambulance, which prompted a further stretch of the wheelbase …

… to 135in (3,429mm), as seen on this Wadham Stringer-bodied ambulance.

to provide the basis of an ultra-long ambulance for use in remote mountain areas. The vehicle was evaluated at Solihull and it persuaded Special Projects to look at building an ultra-long wheelbase chassis for ambulance use.

So the Department developed a 35in (890mm) chassis extension for ambulance use, and once again gave the conversion contract to Spencer Abbott. Wadham Stringer and Herbert Lomas both designed ambulance bodies for the new 135in (3,429mm) wheelbase chassis, the first Wadham Stringer examples appearing in 1976. These were very long vehicles, having not only an extra 35 inches in the wheelbase but also the same 10-inch body overhang behind the chassis at the rear to enable folding access steps to be fitted. Although they did have Land Rover Approval, the ultra-long wheelbase and extended rear overhang detracted quite seriously from the off-road ability of these vehicles. The 135-inch ambulances were always less numerous than the 110-inch types, and were for export only, although one Lomas example did remain in the UK.

FIRE TENDERS

Solihull built its first two fire tender conversions over the summer of 1970, registering them as ELA 831 J and ELA 832 J. Both were straightforward adaptations of the standard vehicle, with first-aid water tanks, Sun fire pumps and ladders in a rack on the roof. However, these were once again unsatisfactory vehicles. The position of the fire pump at the rear made access to the centre of the body very difficult, with the result that equipment carried on board was as good as inaccessible.

Just as with the prototype ambulance conversion, the real difficulty was that the standard Range Rover was too small. Unfortunately, the 110-inch chassis which had solved the space problem for the

The first two fire tender prototypes were based on very early standard vehicles. They were pictured at Birmingham Airport in September 1970.

ambulances was simply not strong enough for the much greater payloads normally carried by a fire tender. So Special Projects turned for assistance to Carmichael of Worcester, who were already approved converters of Land Rovers into fire appliances.

There was little point in turning the Range Rover into a direct competitor for the light fire tender business where the long-wheelbase Land Rover was already well-established. Instead, Special Projects and Carmichael drew up what they called a Rapid Intervention Vehicle, essentially a high-speed first-response airport crash tender. To give the necessary additional space, the chassis was stretched by 40in (1,016mm) behind the rear axle, and the additional length was supported by an undriven third axle. The body was cut behind the B-pillar and removed, and in its place Carmichael built an aluminium alloy structure, containing equipment racks at the side and pumping equipment at the rear. There were ladder racks on the roof and the first-aid foam tank was carried amidships to give even weight distribution.

The prototype was converted from YVB 152 H (chassis number 3550002A), and the conversion was completed by April 1972. It proved to be a very successful vehicle, and was granted Land Rover Approval. Carmichael marketed it as the Commando, and it remained in demand as an airport fire tender all over the world until the mid-1990s. Towards the end of the 1970s, Carmichael also made their three-axle chassis available for custom and leisure vehicles (*see* Chapter 8).

The prototype Carmichael Commando had an aluminium alloy body, but production examples built by the company had GRP panelling for the upper rear bodywork. All the earliest vehicles used the long doors

Carmichael's Commando six-wheel fire tender was a very successful adaptation of the Range Rover. This one dates from around 1980 and was based on the two-door model.

The four-door Range Rover formed the basis of most Carmichael Commandos built in the 1980s. The majority had a roofline like that used on Commandos for the British military, but a very small number of early ones had this style of roof, like that on the two-door types.

The British Armed Forces designed their own Rapid Intervention Vehicles. All of them were built on Carmichael chassis, but there were different suppliers of bodies and equipment. This is an early 1980s example, bodied by Gloster Saro.

of the two-door Range Rover, but after 1981 Carmichael made available a version based on the four-door Range Rover, which offered room for three extra crew members in the cab.

Nevertheless, a version of the fire tender with four doors was built in quite large numbers before Solihull announced its own four-door body. In the late 1970s, the Royal Air Force wanted a batch of fire tenders with four-door cabs, but Carmichael were unable to oblige. So the RAF put the contract out to tender, and ended up buying Carmichael 6 × 4 chassis which they then had bodied to their own design by Gloster Saro. These vehicles had a distinctive cab configuration, with a high-mounted pull handle on each rear door. A later contract was won by HCB-Angus, who supplied vehicles with the same special four-door cab. Thereafter, however, RAF deliveries used the doors of the standard four-door Range Rover and were supplied in batches by both Carmichael and Gloster Saro.

RESCUE VEHICLES

In Germany, the Range Rover was adapted as a fast-response vehicle for dealing with serious road accidents. At least three versions appear to have been available. The earliest appeared in the late 1970s, marketed by British Leyland importer, A. Bruggemann and Co. GmbH. What they

The German firm of Brinck was one of several which converted Range Rovers into emergency rescue tenders for motorway incidents. This is an early 1980s example.

called the *Schnellbergungswagen* (Rapid Rescue Vehicle) was essentially a two-door Range Rover kitted out with cutting gear, extinguishers and other items which would help in getting injured occupants out of a wrecked vehicle. The conversion's development was allegedly sponsored by a weathy businessman whose son had died after an accident when appropriate equipment did not reach the scene in time.

The other two versions were both known as *Vorausrüstwagen* (Fast Response Vehicle) conversions, and were made by fire engine specialists. Bachert of Bad Friedrichshall had a four-door version on the market by 1981 or 1982, and Brinck of Münster introduced a similar vehicle in 1984. Both used the standard wheelbase and standard bodyshell.

THE MILITARY RANGE ROVER

At the same time as Special Projects were looking at Police, ambulance and fire tender conversions of the Range Rover, so the Government Sales department at Solihull were wondering about the vehicle's potential military uses. It was unlikely to appeal as a load-carrier or troop-carrier, but there were certain specialist roles which it might usefully fulfil. Prototype 100/3 (AGN 316 G) was therefore kitted out as a communications vehicle, with the rear seats removed and replaced by racks to carry radio equipment. The vehicle was probably little more than a mock-up, but no doubt Government Sales envisaged fitting a 24-volt electrical system like that on some military Land Rovers.

A second vehicle was prepared and shown at the 1970 Commercial Motor Show, but Solihull hedged its bets. This vehicle also had radio equipment, but was described as a Range Rover Military Command Vehicle. The handout given to interested customers read:

> It must be emphasized that the Range Rover is an extremely versatile vehicle and can be adapted for many special applications. In its present form as a command vehicle, the Range Rover only serves as a basis for discussion, on which potential customers are invited to express their views on how it might be adapted for their own particular requirements.

In fact, the Range Rover never was a big success as a military vehicle. The British armed forces took only small numbers, some allegedly going to the SAS Regiment for specialist duties and a few armoured examples being used by the British Army as transport for senior commanders in volatile trouble-spots. Both the Army and the Royal Navy also took delivery of Wadham Stringer-bodied 110-inch ambulances, beginning in 1977 and continuing into the early 1980s.

Nevertheless, the converters (*see* Chapter 8) did their best to interest military and paramilitary forces in specially-adapted Range Rovers. Wood and Pickett, more famous for their luxurious custom conversions, marketed what they called the Aintree Multi-Role vehicle, and this found favour in some Middle Eastern states as a troop carrier, particularly to carry the personal bodyguard of a ruler or other VIP. In essence, it was a pick-up conversion of the two-door Range Rover, with inward-facing bench seats in the rear which could carry four troops between them. Although marketed by Wood and Pickett, this vehicle had actually been designed by Ogle Design of Letchworth.

FLM Panelcraft also built some open personnel carriers with folding windscreens, and among their users were the Royal

During the Gulf crisis in early 1991, local civilian Range Rovers were commandeered by the armed forces and were highly valued for their comfort and versatility. This one wears the inverted V symbol of the Coalition forces.

The Wood and Pickett Aintree Multi-Role vehicle was a Range Rover conversion designed for military and paramilitary forces and it found some orders overseas.

Guard in Saudi Arabia. A E Smith built some interesting paramilitary CCTV Range Rovers for surveillance duties in Mecca, and there were other special military vehicles – some apparently designed more for parade work than for combat duties – by Glenfrome and Townley.

There were even three-axle military Range Rovers on offer, although it is far from clear whether they found any takers. Publicity material from Scottorn Trailers contained artist's impressions of a three-axle military Communications Vehicle (based on an estate body) and a Reconnaissance Vehicle (with an open cab and open rear body containing a cannon on a swivel-mount), but no pictures of real vehicles have ever come to light. If such vehicles were built, they would have used the three-axle chassis which Scottorn developed in conjunction with Land Rover in 1978. Unlike the Carmichael three-axle

FLM (Panelcraft) built this special Range Rover in the mid-1970s for the Royal Guard in Abu Dhabi.

fire tender chassis, this one had drive to all six wheels, achieved by means of a purpose-built 'through-drive' second axle. Other three-axle military Range Rovers were offered by SMC Engineering of Bristol, and customers could choose from 6 × 4 or 6 × 6 drive configurations. However, once again, no pictures of actual vehicles have been found.

VANS AND PICK-UPS

In some overseas territories, import taxes on luxury vehicles like the Range Rover made the showroom price unacceptably high. However, Solihull soon discovered,

This remarkable machine was built by Glenfrome for Saudi Arabia on a 135in (3,429mm) wheelbase. With automatic transmission and a matching red interior, it was clearly intended for parade duties rather than for a serious military role.

A Glenfrome reconnaissance vehicle for a Middle Eastern customer. The dropside configuration was used on similar vehicles by Wood and Pickett.

Very few Range Rover vans were sold in the UK, but this one spent many years with the Metropolitan Police Bomb Squad. After being withdrawn from active service, it was transferred to the force's off-road driver training school.

with the aid of its overseas dealers, that vans were not subject to import tax. Some customers were quite prepared to buy a two-door Range Rover as a van, avoid the import tax and then (at their own risk) convert the vehicle into a 'proper' Range Rover once it was safely in their hands. So for them, the Range Rover van was created.

The prototype Range Rover van appears to have been built in November 1971. It had ribbed metal panels in place of its rear side windows, no rear seat, and a half-height bulkhead behind the front seats. Production versions had plain metal panels in the rear side window apertures, but otherwise there appear to have been no changes. In Denmark, different regulations allowed the Range Rover van to have fixed glass in its rear side windows, and in France (where it was called the Range Rover Utilitaire) the standard sliding windows were retained; but in all countries it was delivered without a rear seat and most examples also had the half-height bulkhead, which was made of wood. Vans continued to be made into the 1990 and in fact the very last two-door Range Rover built, in January 1994, was a van destined for Portugal.

It might have been logical to expect Solihull to experiment with a pick-up version of the Range Rover, too, as such a vehicle would have been able to provide ideal support for shooting parties. However, no such vehicle seems to have been considered, although one very stylish Range Rover pick-up was built in 1971 or 1972, allegedly to special order for HRH Prince Philip.

It would be 1981 before a pick-up version of the Range Rover was seriously considered, and the idea came from Carbodies of Coventry (makers of London's black taxicabs), who had just begun to manufacture some body panels for the new four-door Range Rover.

Carbodies' plan appears to have been to create what was effectively a luxury Land Rover, by combining the power and comfort of the Range Rover with some of the Land Rover's versatility. As the Unitruck sales brochure later put it, 'the Range Rover's proven ability to offer comfortable accommodation in off-road conditions [is] a prime advantage over more agricultural vehicles'.

The idea made sense, and the Unitruck's combination of the best elements of Land Rover and Range Rover should have been a winner.

The Unitruck was announced some time around April 1982. It was based on a two-door Range Rover to keep costs down, with the rear of the body removed at waist level to create a pick-up configuration. However, the rear seat remained in place and was incorporated into a new passenger cabin to give a five-seater 'crew cab'. To leave as much room as possible in the pick-up bed, the spare wheel was relocated above the cab roof on a structure not unlike a rollover bar. This structure certainly gave the Unitruck a unique appearance, but it can hardly have made the business of changing a wheel very easy. A bonnet-mounted spare would surely have been a better solution, unless Carbodies were concerned that it reduced forward visibility too much for off-road driving.

The Unitruck was certainly intended to be used off the road: Carbodies expected their customers to be farmers, who would use the vehicle for spraying, pellet spreading and other general purpose farm duties. Well aware that farmers often prefer diesel fuel to save costs, and handicapped by the lack of a factory-option diesel engine for the Range Rover, they had therefore ensured that the Unitruck could be cheaper to run than an ordinary Range Rover. On the options list was an LPG conversion, which would allow the vehicle to run on cheap liquid petroleum gas and go back to petrol at the flick of a switch when

The Unitruck conversion by Carbodies failed to find a market in the early 1980s.

required. (In the early 1980s, it looked as if LPG might have a future in Britain, but the buying public was deeply suspicious of the new fuel and it never caught on.)

It was Carbodies' misfortune to introduce their new vehicle at precisely the wrong time. The new coil-sprung Land Rover One Ten was announced some six months after the Unitruck appeared at the 1982 Motor Show and it offered all the traditional versatility of the Land Rover in tandem with a Range Rover-derived coil-spring suspension. In addition, it had an uprated V8 engine which gave it more than adequate performance. In the face of that, the advantages of the Unitruck seemed to evaporate, and the matter of pricing put the final nail in its coffin: the Unitruck cost half as much again as a V8-powered One Ten pick-up.

So the Unitruck never did go into series production. Only around five examples were ever made, all of them prototypes. Four are known for certain. LJT 276 P was probably the very first vehicle, clearly converted from a second-hand Range Rover of circa 1976 vintage. Then there was SWT 824 W, registered in Yorkshire rather than Carbodies' Coventry home and therefore probably bought in as a one-year-old to make the second prototype. The third example known was WAC 434 X, its Solihull registration suggesting that the vehicle may have come from Land Rover, who did in fact grant full Approval to the Unitruck conversion; this one may have been the 1982 Motor Show vehicle. Lastly, JSF 941 X was bought by Alka-Seltzer and became a support truck for their advertising hot-air balloon.

7 Conversions: High-Performance Range Rovers

Although Range Rovers have always been very good road cars, far better than such big 4 × 4s had any right to be, standard versions have never had the performance or the road manners of saloon cars in their price bracket. To the majority of buyers, that has not been a problem; they have been more than happy to own a vehicle which has such a competent blend of off-road and on-road ability.

However, from the earliest days, most customers have bought Range Rovers purely as road cars and the majority of examples have probably never been further off-road than a muddy field at a gymkhana. These vehicles have been bought as a direct substitute for a saloon car of similar cost and it is hardly surprising that some of their owners have wondered whether saloon car road performance might be available.

Towards the end of the 1970s, a small number of companies therefore began to offer conversions to make the Range Rover a better road car. It had always been possible to improve performance simply by changing the carburettors and camshaft or by increasing the cubic capacity of the engine, but these conversions went very much further than that. For a start, they were properly engineered; and in addition they were fully warranted. This sort of peace of mind was not cheap to buy, but the small numbers of wealthy buyers who decided to invest in converted vehicles rarely regretted their decision.

As the standard Range Rover was improved during the 1980s, so there became less for the conversion specialists to achieve and few of them were prepared to risk the development costs of a faster or better-handling vehicle when there was always the risk that Land Rover might equal or surpass their efforts with the next round of production-line changes. However, by the early 1990s the position had changed. The major new factor was the arrival of the Discovery in 1989, which used what was essentially the Range Rover chassis together with its power train, axles and suspension. It was quite clearly Land Rover's policy to keep Discovery one or two steps behind the Range Rover in terms of mechanical development, in order to safeguard the more expensive vehicle's image and sales. This meant that any conversion developed to improve this year's Range Rover would probably also be suitable for this year's and next year's Discovery, which gave it a much greater sales potential. As a result, a number of new Range Rover conversions were announced in the vehicle's final years.

Many Range Rover enthusiasts who

have never driven these conversions wonder what they are like, particularly off the road. The simple answer to that is that the most competent conversions provide exhilarating road performance without losing the low-speed tractability which is so important for off-roading. Similarly, the best of the suspension modifications improve road handling without compromising the axle articulation needed for heavy off-road work. At worst, these conversions simply need a different set of tyres for heavy off-road use, as the high-performance road tyres many converters recommend have the wrong sort of tread pattern to give grip in mud or sand.

If these conversions really are so good, why have the Range Rover's makers never attempted to emulate them? The answer is simply that they are expensive – enormously so, in some cases – and that even building them in the high volumes associated with the Range Rover's production line would add unacceptably to the showroom price of the vehicle. Most customers have always found the standard vehicle good enough and it has been considerably improved over the years to meet changing expectations. Adding features which most customers can manage without, while increasing the showroom cost of the vehicle, would certainly have lost the Range Rover some sales. As it was, Land Rover managed to satisfy most of the customers most of the time – and that is surely the secret of success in selling any product.

SCHULER PRESSES AND OVERFINCH

The major player in high-performance Range Rover conversions has always been Overfinch, a company which began operations in 1975 under the rather misleading name of Schuler Presses. The company has become to Land Rover what AMG is to Mercedes or Alpina to BMW, the only difference being that the two German car makers have adopted their respective tuning specialists, while Land Rover has not.

The high-performance Range Rover business was the brainchild and pet project of Arthur Silverton, Managing Director of the UK branch of Schuler Presses, a leading German manufacturer of industrial presses. Silverton was able to persuade the parent company that there was a future in high-performance Range Rovers and as a result he was granted a large development budget to explore the possibilities.

Silverton's idea was not only to make a Range Rover accelerate faster and reach higher speeds; his aim was to improve the vehicle from top to bottom, giving it better handling, better braking, and eliminating all the shortcomings of its rather agricultural four-speed manual transmission. As an initial stage, he tried out the 192bhp Australian 4.4-litre version of the V8 engine (from the Leyland P76 saloon), but later switched his attention to a high-output version of the 3.5-litre V8 engine, a whole new transmission and an improved braking system.

The transmission used an American Chrysler A727 automatic gearbox (the same one which would later be adopted for the production Range Rover Automatic) and a Schuler–Voith chain-driven transfer box which incorporated a viscous-coupled centre differential (similar in principle to the Borg Warner transfer box adopted for production some ten years later). However, the key ingredient in the Schuler Range Rover was a Ferguson Formula anti-skid brake system, operated by a sensor on the transfer box output shaft. While it was not as sophisticated as the Wabco ABS developed

Demonstrating that its axle articulation is every bit as good as the standard model's, this is a Schuler 350HE, dating from around 1980. Note the twin exhausts.

for production vehicles a dozen years later, it was technically very advanced for its time and was also very effective.

By 1980, Schuler were ready to go into business with their 350HE conversion, which gave the standard Range Rover a road performance broadly similar to that of the fuel-injected models which Land Rover announced five years later. The automatic transmission, Schuler–Voith transfer box and anti-skid brakes were optional but recommended, and the 350HE conversion was immediately popular. Although no precise figures exist, the present owners of the company estimate that around 100 vehicles were converted to 350HE specification each year it was available.

Meanwhile, however, Silverton was working on an altogether more ambitious project. Instead of simply uprating the Range Rover's 3.5-litre V8 engine, he had fitted a development vehicle with a 250bhp 5.3-litre Jaguar V12 engine. As this engine was considerably longer than the V8, the Range Rover had been stretched to accommodate it, all the extra length going in between windscreen and front axle. Performance was allegedly exhilarating, but ultimately the cost of the 'stretch' caused the programme to be abandoned.

One reason for choosing the Jaguar V12 engine had been that it offered a very large power increase and was reasonably cheap to buy. The only other engines which met

The remarkable Schuler V12 prototype had a 5.3-litre Jaguar engine under its lengthened bonnet, but the cost of the stretch prevented it from going into production.

The 5.7-litre GM engine was such a good fit in the Range Rover's engine bay that it might have been designed for the vehicle. This is the very first example which Schuler fitted to a client's vehicle; it went to the Middle East.

those same criteria were American V8s and Schuler now turned to the Chevrolet 350 cubic inch (5.7-litre) engine, as used in various saloons and sports cars. This was an altogether better choice, fitting under the Range Rover's bonnet without surgery to the vehicle; and Chevrolet were also prepared to build engines incorporating certain features to Schuler's specification.

The Schuler 570T was made available to the public in 1982, its name reflecting the engine's capacity in litres and the 'T' standing for torque – all 280lb/ft of it, developed at 1,800rpm. The 205bhp carburettor engine gave the 570T a maximum speed of 117mph (188km/h) and a 0–60mph acceleration time of 9.5 seconds, driving through the GM400 automatic gearbox which came as a standard part of the conversion. This was performance which the standard Range Rover was not equipped to handle, and as a result the 570T invariably came with the Schuler transfer box, anti-skid brakes and ventilated front brake discs. An extra-cost option was a handling package which affected springs and dampers, and provided different suspension bushes in order to improve stability in corners and reduce the Range Rover's characteristic body roll.

The 570T proved even more popular than the 350HE conversion had done and it appears that many owners of 350HEs brought them back to Schuler to have them uprated to 570T specification! The cost of the full 570T conversion was formidable – something like two-thirds of the cost of a brand-new Range Rover – but it did turn the vehicle into a true high-performance machine. Off-road ability was unaffected (although Schuler recommended fitting the high-performance road tyres developed for Rolls-Royce in preference to the standard dual-purpose type) and even fuel consumption was only marginally worse than for a standard Range Rover of the day, at around 15.5mpg (18.3l/100km). Not surprisingly, some of the 570T conversions were carried out under contract to companies like Wood and Pickett as part of a full customization for Middle Eastern clients.

With an excellent reputation already established, the Range Rover division of Schuler changed its name to Overfinch in 1985. The parent company had become slightly embarrassed by their close association with the Range Rover, not least because they were supplying industrial presses to Mercedes-Benz, makers of the rival G-Wagen; and as a result, Arthur Silverton established Overfinch as a separate concern. Over the next few years, he continued to market the 570T conversion with great success and continued to work on further Range Rover refinements. In 1986, the 570T was supplemented by a 570S model (the 'S' stood for Sport), which had a GM 700 four-speed automatic transmission to handle increased engine power and torque. However, by 1988 Silverton was keen to retire and so Overfinch was sold to new owners.

Under the new trading name Overfinch 2, the new owners expanded the company's range of activities and gradually gave it a higher profile. They retained the 5.7-litre conversion and in 1988 introduced a 5-litre conversion at a lower price. Once again a GM engine, this was offered in the Overfinch 500E and 500S models. The following year saw a more powerful edition of the 5.7-litre engine in the 570S2, and then in 1990 fuel injection replaced carburettors in Overfinch's conversions as Chevrolet made the changeover on the production lines. Model names changed to reflect this, becoming 500i and 570TPi.

A Precision Steering conversion introduced in 1990 enhanced chassis dynamics still further, and a quick-ratio speed proportional steering option announced the following year gave even greater responsiveness. That year (1991), the 570Ti (Torque injection) conversion was announced, with an uprated version of the Range Rover's standard ZF automatic box instead of a GM gearbox. As most vehicles provided for conversion were relatively new and therefore already fitted with the Borg Warner chain-driven transfer box and Wabco ABS, demand for the Schuler–Voith transfer box and Ferguson Formula anti-skid brakes gradually ceased, and Overfinch concentrated on engine, suspension and steering work. In response to customer demand, however, they introduced various cosmetic alterations during 1993–1994, including Recaro sports seats, retrimmed interiors, and unique 8 × 16 alloy wheels.

Overfinch conversions are widely regarded as the best of them all, retaining the essential balance of the original vehicle but endowing it with sports-saloon acceleration matched by handling and braking which ensure that everything remains perfectly under control in corners or when an emergency looms up ahead. There is nothing quite like cruising at a relaxed 100mph (where the law allows it) in a big and luxurious vehicle and then feeling the acceleration which is still available when the throttle is floored.

JANSPEED

Jan Odor's Salisbury-based Janspeed Engineering developed a reputation for improving performance with special exhaust systems and turbochargers towards the end of the 1970s, and not surprisingly it developed a conversion for the Range Rover. The original conversion used a single Roto Master turbocharger, but this was later changed for two smaller Roto Master RM60s in order to reduce turbo lag. An intercooler was also fitted.

With this later conversion, the 3.5-litre carburettor engine put out 220bhp at 4,200rpm and 250lb/ft at 2,700rpm, enough

Extensive additional plumbing in the engine bay marks this out as a Janspeed twin-turbo conversion of a carburetted Range Rover.

to take even a sluggish three-speed automatic model to 103mph when *Autocar* tested one in June 1985. The same magazine recorded the 0–60mph acceleration as 12.3 seconds, some 4.5 seconds faster than the standard automatic could achieve. Some UK Police Forces are said to have used Janspeed turbocharged Range Rovers for motorway patrol work and the Janspeed conversion was allegedly also fitted to a number of armoured vehicles used for VIP protection duties, to help overcome the extra weight of the armour plate and armoured glass.

When the fuel-injected models arrived in the mid-1980s, Janspeed developed a single turbocharger conversion to suit them, using a Roto Master T4 turbocharger with an intercooler and a differently programmed engine management computer. In this guise, Janspeed claimed 238bhp at 4,500rpm and 298lb/ft of torque at 4,000rpm, with a 0–60mph time of under 9 seconds and a maximum speed of 125mph (200km/h). To go with this vastly improved performance, the company also marketed a handling package.

JE MOTORS: THE DAKAR RANGE ROVER

JE Motors (latterly John Eales Development) of Coventry has been the leading Rover V8 tuning specialist since the late 1970s, when the company developed a 3.9-litre version of the engine which sports car maker TVR later adopted for its 390i model. Characteristic of all the JE Motors engine conversions has been a concentration on bottom-end torque rather than top-end power, and this has made them ideally suited to a heavy vehicle like the Range Rover.

In the early 1980s, JE Motors developed a 4.2-litre V8, and a few years later a 4.5-litre type, both of which went into a number of Range Rovers built to customer specifications by companies catering for the Middle Eastern custom-building trade. The company also forged close links with Land Rover at Solihull, being asked to develop and build the engines for the 1987 and 1988 factory-backed Paris–Dakar Range Rovers; different versions of these engines also went into other Range Rovers competing in the 1988 Paris–Dakar Rally. It was as a result of this work that JE Motors brought to the market its own Dakar Range Rover conversion in the late 1980s.

The centrepiece of the Dakar Range Rover was a 4.2-litre engine with 250bhp at 5,500rpm and 280lb/ft of torque at 4,000rpm. This gave excellent mid-range and high-speed acceleration, as well as a 0–60mph time of less than 10 seconds. The JE Motors Dakar conversion also had uprated braking and suspension modifications to keep this extra accelerative ability and high-speed performance under safe control.

TURBO TECHNICS

When the conversion specialists began to show renewed interest in the Range Rover in the early 1990s, quickest off the mark – at least, in terms of bringing a product to market – were Turbo Technics of Northampton. This was a company which had already established a name for itself in the field of high-performance saloon conversions.

The Turbo Technics Range Rover conversion was introduced at the October 1991 London Motorfair and consisted purely of an engine conversion. No handling or braking enhancements came as part of the package, although Turbo Technics did recommend the Warwick Banks and Harvey Bailey handling kits. Conversions were available for both fuel-injected 3.5-litre and 3.9-litre versions of the engine.

The essence of this conversion was a Garrett T3 turbocharger. The engine's compression ratio was lowered by means of a special head gasket and the combustion chambers were enlarged, while there was a supplementary electronic control unit for the fuel injection system. In addition, a large-bore stainless steel exhaust was added. Turbo Technics claimed 258bhp at 5,000rpm for the turbocharged 3.9-litre engine, together with 330lb/ft of torque at 2,500rpm. They made no claims about maximum speed, but quoted a 0–60mph

Under the bonnet of a Turbo Technics Range Rover, with a 258bhp turbocharged 3.9-litre edition of the V8 engine.

acceleration time of 8.1 seconds and impressive in-gear acceleration figures for an automatic Vogue SE model: the quoted 50–70mph (80–110km/h) figure of 6.5 seconds was around a second faster than the standard model could manage.

This was a cost-effective conversion, giving smooth and powerful acceleration once the turbocharger made its presence felt at about 2,000rpm. Mid-range and high-speed acceleration were noticeably better than with the factory-issue 3.9-litre engine and this was an altogether useable performance enhancement for the Range Rover.

TWR

One of the new entrants to the Range Rover market in this period was Tom Walkinshaw Racing, a company which was no stranger to the Rover V8 as it had been closely involved with the works Rover Vitesse saloon racers in the first half of the 1980s. As its first brush with the Range Rover, however, TWR designed the star-pattern alloy wheels and the bodykit which became available through Land Rover Parts in 1992. The following year, it began to market certain bodykit additions which Land Rover had turned down; and towards the end of the year it introduced its 4.7-litre conversion.

The 4.7-litre (4,651cc) engine was developed as an enhancement for the 3.9-litre and 4.2-litre types available from the factory, and shared their bore size. However, its long-stroke crankshaft was used to boost the mid-range torque and it was TWR's

TWR's own 4.7-litre demonstrator was tricked out with a full body kit.

proud boast that the engine exceeded the 230lb/ft of the factory-issue 3.9-litre engine all the way from 1,500 to 4,900rpm. On the road, the additional torque was very noticeable, and it increased the engine's flexibility at all speeds. Power was limited to 230bhp, mainly because TWR believed that owners would simply not wish to drive at more than 120mph (195km/h) for long periods – even in those few countries where such speeds were legal.

To match the extra performance, TWR also offered a handling package. This consisted of Avon Turbospeed 255/60 tyres (from a Bentley Turbo R) on the standard 16-inch wheels, a tougher steering damper to control feedback and firmer progressive-rate coil springs with adjustable Koni dampers. This combination gave much flatter cornering than the standard vehicle offered, with a precision about the steering which was enough to frighten anyone used to the production vehicle's tendency to run wide on corners. Drawbacks were a slightly harder ride and a slight loss of off-road ability caused by the road tyres.

OSELLI

Oselli Engineering of Oxford established its reputation by power-tuning standard road engines, and among these was the ubiquitous Rover V8, on which the company started working in the late 1970s. Four-wheel drives had never been part of Oselli's repertoire, but some time around 1992 it decided to look at what might be done with the Rover V8 to improve the road performance of the Range Rover. Seeking a niche in the market, it turned its attention to the Range Rover LSE and its 4.2-litre version of the V8 engine.

The result was a conversion known as the Oselli Challenger, which boosted the 4.2-litre engine's power to 230bhp at 5,300rpm and increased its torque to 270lb/ft at 3,500rpm (hence its alternative name of the 270PF conversion). These figures were achieved by modifying the induction system, changing the camshaft, fitting uprated hydraulic lifters and – most important of all – rechipping the engine management ECU. To make the best of these changes, Oselli also fitted new large-bore exhaust manifolds and a large-bore exhaust system.

Maximum speed was a claimed 120mph (193km/h), although speed was limited to 110mph (177km/h) on standard road tyres; full performance demanded an expensive change to the Avon 255/60 VT 16 tyres designed for the Bentley Turbo R. Acceleration from rest was unexpectedly eager, and despite a torque peak some 250rpm higher than in the standard vehicle, the broad spread of torque gave the Challenger improved acceleration at all speeds.

The Challenger was announced in 1994, and was followed by three even more powerful conversions. These were the Chieftain, the Conqueror and the Charger. The Conqueror conversion used an Opcon Autorotor supercharger to boost the power of the 4.2-litre V8 to 330bhp at 5,400rpm and the torque to 350lb/ft at 4,350rpm. The Charger, meanwhile, took the engine out to 4.8 litres and added a supercharger as well. In all cases, large-bore manifolds and exhaust system were added, and Oselli also offered suspension changes to match the handling to the vastly increased performance.

LICHFIELD

Lichfield conversions were announced early in 1993. Their name came from Lord Lichfield, Royal photographer and an

Lichfield conversions like these used uprated V8 engines prepared by TVR Power, a division of the sports car company.

enthusiastic Range Rover owner, but the company which built them was known as TRM. TRM were based in Melton Mowbray and were associated with the franchised Land Rover dealership known as TMS. Formerly known as Tractamotors, TRM had offered aftermarket performance conversions to Land Rover products since the late 1980s.

Not every Lichfield conversion was designed for high performance, because TMS also offered a range of interior and other cosmetic enhancements. However, also on the options list (and the only vehicles fitted with the blue oval Lichfield badge) were high-performance derivatives of the Rover V8 engine built by TVR Power in Coventry, a division of the Blackpool-based sports car manufacturer TVR.

Both 4.5-litre and 5-litre versions were available, the latter with 278bhp at 4,500rpm and torque of 340lb/ft at 3,250rpm. The 5-litre offered electifying road performance, but for safety reasons was only available with a handling kit consisting of new springs, dampers and a stiffer rear anti-roll bar; Bentley Turbo R tyres were also part of the package. In addition, TMS offered a steering enhancement, consisting of a new drop arm, track rods and damper, which sharpened up the steering response considerably.

During 1994, TMS also introduced a performance enhancement package for the 300Tdi turbodiesel engine. This was intended primarily for the Discovery, but could also be fitted to Tdi-engined Range Rovers.

8 Conversions: Custom-Built Range Rovers

The performance and comfort of the Range Rover made it an ideal basis for a large number of special conversions intended, like the standard vehicle, to carry passengers. On the whole, Solihull stayed well clear of this market, fearing that it would be volatile and recognizing that a company committed to mass-production could not easily cope with building vehicles to individual customer specifications. A few vehicles certainly were specially turned out for VIPs, including one for Prince Rainier of Monaco in 1971 and a pair of hunting vehicles for Prince Abdullah of Saudi Arabia in 1973. However, it was not until 1993 and the arrival of the Autobiography service from Land Rover Special Vehicles that Solihull even began to explore the wider possibilities of this market. In the meantime, a number of specialist coachbuilders adapted the Range Rover in many different ways, sometimes with the benefit of Land Rover Approval but often without it.

THE RANGE ROVER CAMPER

In their declining years, many former Range Rover ambulances have been turned into mobile caravans by enthusiastic DIY owners. However, the basic Range Rover was always too small to make a wholly satisfactory camper, as specialists Carawagon International of Sunbury-on-Thames discovered.

Carawagon already had a thriving business in turning Land Rovers into campers, and when the Range Rover was announced they decided to see if they could carry out a conversion along similar lines. The prototype Carawagon Range Rover was built from a very early vehicle registered EPK800J, and was examined by the Land Rover engineers at Solihull at the beginning of March 1971. It was then offered for road test to *Autocar*, whose Stuart Bladon was cautiously enthusiastic about it in the magazine's 24 June 1971 issue.

Lack of space inside the Range Rover made this two-berth conversion rather less appealing than the Carawagon campers based on a long-wheelbase Land Rover. Although the two vehicles had the same

The Carawagon conversion was rare. This one had been well used by the time it was pictured in the 1990s.

interior width of 64in (1,625mm), the Range Rover's floor area was seven inches shorter than the 109's, and the internally-stowed spare wheel stole valuable space. Carawagon were also unable to fit a full-size version of their famous elevating roof, so that maximum standing height in the Range Rover was just 64in (1,625mm), as compared to 74in (1,880mm) in the 109 Land Rover.

Nevertheless, the Range Rover conversion had been executed with Carawagon's customary ingenuity. Central to it was a large 'magic box' between the front seats, which incorporated storage space for cutlery, crockery and saucepans. The lid of this box contained the stove, and folded forwards to rest on the dashboard when in use, revealing a basin and electrically pumped water supply underneath. Finally, a cantilevered table pulled out towards the rear. The rear seat had been removed and replaced by inward-facing seats on either side, which folded down to make two beds. There was also a wardrobe in the back, opposite the spare wheel, water storage to the rear of that, and an assortment of pouches and pockets for additional storage.

Very few of these campers seem to have been built. Lack of space must have been one reason; cost another. At £3,040 without extras in August 1972, the two-door Carawagon Range Rover was hardly a bargain as compared to £2,450 for a six-cylinder 109-based Continental Ultimate, complete with four berths and appreciably more living space. The Carawagon Range Rover was still listed in Land Rover's own Range Rover Special Conversions catalogue as late as 1980, but by then Carawagon themselves were more interested in promoting a cheaper version called the Safari Range Rover. This did not have the expensive elevating roof or the full interior conversion, but instead came with two inflatable mattresses for the rear load area, special seats, window blinds and zipped clothes storage bags. Once again, the vehicle seems to have enjoyed only limited success.

ARMOURED RANGE ROVERS

'We live in a world of increasing violence and, for many key executives, diplomats, Heads of State and other public figures, the possibility of personal attack during a road journey or ceremonial drive cannot be ignored.' Thus did Land Rover explain the need for an armoured Range Rover in a sales brochure dating from 1986. In fact, the Range Rover was in many ways the ideal basis for an armoured limousine. It was sufficiently prestigious and stylish to be entirely acceptable as VIP transport, and it was sufficiently strong to bear the additional weight of armour plate.

The whole point of armour-plated vehicles for VIP protection is that they should be unobtrusive, and for security reasons no manufacturer will provide details of end-users or of the precise degree of ballistic protection fitted to any individual vehicle. In addition, the specification varies from vehicle to vehicle, and is dependent on the nature of the threat to the VIP using it. It is therefore impossible to uncover precise information about the armoured Range Rovers which were built, but it is possible to establish a general outline.

The first armoured Range Rovers were probably built in the late 1970s, and it appears that some of them were turbocharged by Janspeed to restore performance lost to their extra weight. Overfinch (*see* Chapter 7) also provided suspension conversions for some MoD Police vehicles in this period, in order to prevent the extra weight of the armour plate from causing handling problems. The identity of all the companies who offered armoured Range Rover conversions is not clear, but Hotspur Armoured Products of Neath in West Glamorgan were certainly among them.

In 1985, Special Vehicle Operations at Solihull announced their Discreetly Armoured Range Rover, with ballistic protection and other special features by specialists MacNeillie. Among other items on offer were a fixed division behind the front seats, pull-down face masks linked to an emergency oxygen system, an electronic tamper-detection system, run-flat tyres and a communications system which enabled those inside the vehicle to speak to those outside without opening a window. These vehicles were fitted with anti-roll bars (five years or more before the 'civilian' version became standard on all Range Rovers) and had no self-levelling strut at the rear.

Many armour-plated Range Rovers had twin side-hinged rear doors instead of the standard tailgate, although Land Rover's own Discreetly Armoured conversion did not. Generally, the wheels were plain discs (often with run-flat inserts and special tyres), and in some cases they were fitted with a moulded trim to make them look like standard Range Rover Rostyle wheels.

EXTRA DOORS, EXTRA INCHES

Solihull's Sales Division had not been wrong in anticipating that customers would demand a four-door Range Rover, and as early as 1972 a prototype was built as a feasibility study. Lack of capital and lack of production capacity prevented that from becoming a production reality, but these factors could not stave off the demand. During 1974, the coachbuilders FLM (Panelcraft) of London, who had long-standing links with the Rover Company and had done prototype work for them, developed a four-door conversion for the Range Rover. Although this was in many ways a rather crude design, with sliding windows in its rear doors, it was granted Land Rover Approval and soon found a ready market in the Middle East, where the Range Rover had only recently been introduced.

Very few of these four-door conversions

An armoured Range Rover of the UNHCR in Bosnia in the early 1990s. Note the distinctive twin rear doors. Steel disc wheels are normal wear, but one has been replaced temporarily by a standard Range Rover wheel and tyre.

found customers in Britain – although the first appears to have been built for the Duke of Roxborough – but the market overseas proved sufficiently large for other coachbuilders to develop similar conversions. Wood and Pickett of London, who were already known as builders of special Minis and other exotica, had developed a four-door conversion by the mid-1970s and used it to expand their business into the custom-building trade for the Middle East. Their example was followed by Panther Westwind of Byfleet, who introduced a four-door conversion some time around 1977. All these coachbuilders later expanded their Range Rover business to include other types of conversion as well.

As Chapter 6 makes clear, Solihull had also anticipated the demand for a long-wheelbase Range Rover with extra seats as early as 1971, when it built a prototype high-roof Station Wagon with Spencer Abbott. That vehicle did not prove to be a success, but a market did materialize for Range Rovers with extra carrying capacity, and a number of coachbuilders developed long-wheelbase conversions to meet the demand. In some cases, these conversions retained the standard number of seats, and the additional interior space was used either as lounging room for those in the rear or as additional load space.

There were several sizes of long-wheelbase stretch. The smallest and most practical was to 109in (2,769mm), although some converters preferred the 110in (2,794mm)

Wood and Pickett lengthened the wheelbase and raised the roof on this early 1980s four-door model.

Lengthened wheelbases were the foundation of the custom Range Rover market. This is a 120in (3,048mm) four-door stretch, dating from 1980 and probably carried out by A E Smith.

This 135in (3,429mm) four-door stretch by A E Smith dates from the early 1980s.

size which Solihull had developed for commercial bodywork. Doubling those 9in or 10in extensions gave wheelbases of 118in ((2,997mm) and 120in (3,048mm), the next sizes up. Above those came 124in (3,150mm) and 133in (3,378mm) wheelbases, and the maximum practical stretch of the two-axle chassis proved to be to 135in (3,429mm). This wheelbase was commonly used with six-door body conversions which incorporated a third row of seats in the centre of the wheelbase.

Most of these long-wheelbase conversions went abroad, the majority to the Middle East. Demand for them dwindled towards the end of the 1980s, and tailed off to almost nothing after Land Rover introduced their own long-wheelbase model in 1992. However, one company was still making long-wheelbase vehicles in small numbers until as late as 1994.

Special Vehicle Operations at Solihull also became involved in the construction of at least two special long-wheelbase models towards the end of the 1980s. The first was on a 120in (3,048mm) wheelbase, and was built as an official car for the Mayor of Solihull. It was finished in Royal Blue and entered service in 1987. The second was built in 1989 on a 124in (3,150mm) wheelbase, and was used by Land Rover for VIP transport. In the early 1990s, Land Rover

Special Vehicles at Solihull became more prepared to take on custom conversion work in the 1990s, and this is a mobile office they built on a 120-inch (3,048mm) wheelbase for an Italian customer. The front seats swivelled to face the table for conferences.

Special Vehicles (as it had by then been renamed) also built a special long-wheelbase Range Rover for an Italian client; this vehicle was blue and its front seats could be swivelled round to face a table located between them and the rear bench.

THE SIX-WHEEL ESTATES

Although the three-axle extended Range Rover chassis was originally developed for fire tenders, by the end of the 1970s it had also been used to create estate cars with additional carrying capacity. The first of these was probably the one commissioned by John Woolfe Racing of Bedford in 1977 as a promotional vehicle for their Wolfrace custom wheels, and registered as RRP484R. This was based on a Carmichael 6x4 chassis, and its body was built by Pilling (Coach Builders) Ltd. The whole vehicle was painted black, and it was equipped with striking chromed alloy wheels and huge Thrush 'sidewinder' exhausts. Although it certainly looked striking, it proved to be both rather underpowered and (initially at least) mechanically unreliable.

Still, the idea of a large-capacity estate caught on, and when Scottorn came to develop their 6x6 chassis in 1978, they

The Wolfrace Range Rover was a very early six-door estate conversion; it had 15-inch wheels instead of the 16-inch size fitted to standard Range Rovers. The front seats with wraparound stereo headrests were supplied by Wood and Pickett and the chromed door mirrors were from a Mercedes saloon.

The treatment of the rearmost side window distinguished the Scottorn six-wheel Range Rover from other types. This one dates from about 1979 and was based on a two-door model.

announced estate-bodied versions as their main models. Although early prototypes had standard Range Rover rear side windows, production vehicles had smaller windows in a thick frame which helped to reinforce the rear of the body. These estates remained available into the early 1980s, although probably none were made after Scottorn were taken over by Reynolds Boughton in 1983.

Carmichael also decided to try their luck in this market, and in 1980 they introduced two different conversions based on their 6x4 fire tender chassis. Both had three rows of seats and a raised rear roofline, the Highlander having just two side doors and the Clansman being converted to four-door configuration. A 136in (3,454mm) wheelbase two-axle version of the Clansman was also announced, but was discontinued shortly after its launch. All these Carmichael conversions seem to have remained quite rare, and probably most of them went for export.

The three-axle market also attracted JNR Motors of Aston and SMC Engineering of Bristol. JNR Motors offered their Reveller range in the late 1970s and early 1980s. These had 6 × 4 chassis with a 136in (3,454mm) wheelbase, and came as two-door or four-door estates, or as five-seater crew-cab pick-ups (of which there was a recovery vehicle version). In addition, one 6x6 chassis was bodied as an ice-cream van and was used on beaches in Wales.

SMC Engineering developed their three-axle conversion as an extension of the work they had been doing on three-axle Land Rovers, and made their Sahara Six chassis available for commercial, military and custom bodywork during the early 1980s. In estate form, it came as a 6 × 4 or a 6 × 6, as Ellgee, Ellgee Executive or Armoured Executive models, and with distinctive bodywork by Longwell Green Coachworks of Bristol. This featured a stepped roof, raised by 8, 10 or 14in (203, 254 or 356mm), and customers had a choice of side-opening or up and over rear doors. Ballistic armour on the Sahara Six Armoured Executive model was by Manor Armour Ltd of Bridgwater in Somerset.

Possibly the only Carmichael Clansman to have remained in the UK, this one was in need of restoration when it was pictured in 1995.

THE CUSTOM MARKET

The basic conversions to four-door, to long-wheelbase estate and to six-wheel estate lay at the heart of the custom-building industry which grew up around the Range Rover in the mid-1970s and flourished until well into the 1990s. Most custom-built Range Rovers were made for Middle Eastern clients, and by their nature were individually specified. Western magazines have often commented adversely on these vehicles out of ignorance, and it is therefore important to put them into their correct perspective.

The oil crisis of 1973–4 was followed by a major increase in the cost of oil, and before long the Middle Eastern suppliers of oil had become very wealthy indeed. In Arab countries, it is the social norm for a wealthy man to show his wealth, and this gives rise to a taste for ostentation which is wholly alien to Western social customs. The Range Rover proved to be a first-rate vehicle for Middle Eastern conditions because of its performance in desert sand, and it was much liked by those who could afford it because it offered much greater comfort than any other four-wheel drive vehicle available in the 1970s. For a wealthy Arab, it was thus a natural step to buy a Range Rover and to have it customized to suit his needs and demonstrate his wealth.

Conditions in the Middle East once again dictated the types of conversions which became popular. Large families are common, and so Range Rovers with extra seats were in demand. Many wealthy Arabs enjoy the sport of falconry in the desert, where it is vital for the falconer to be able to follow the fast-moving bird as it flies across the dunes in pursuit of its prey; consequently, many customers ordered vehicles with elevating 'hunting seats' which would enable them to keep field-glasses trained on a bird while their drivers attempted to keep up with it. In later years, some of these

Wood and Pickett's Newmarket Hunter was a conversion which could be used for a variety of desert sports. Note the repositioned fuel filler.

Hunting seats on hydraulic lifts were often demanded by Middle Eastern customers. Here they are in the raised position on a 1980 four-door by FLM (Panelcraft).

vehicles were also fitted with radar tracking equipment for use if the falconer lost visual contact with his valuable bird.

Long-distance desert travel is also common, and so many customers asked for long-range fuel tanks, refrigerated water containers, air conditioning, radio-telephones, and other equipment designed to make such journeys more comfortable. And, of course, all the extra equipment they wanted on their vehicles tended to make them very heavy, so they wanted more powerful engines to compensate. Colour schemes and interior decor which seem bizarre to Western tastes were merely the finishing touches to conversions which were commissioned primarily to serve practical purposes. Perhaps only the convertible Range Rovers qualified as examples of pure ostentation.

The great variety of customer requirements and the large number of companies competing for a finite market inevitably led to some problems, but before long the situation had sorted itself out. The major players in the customized Range Rover market tended to sub-contract elements of their conversions to a network of smaller companies which specialized in certain areas. Thus, a vehicle sold with the badging of one of the major players might actually have a wheelbase stretch and six-door conversion by one company, an engine conversion by a second, an interior retrim by a third company and a whole collection of other items bought in from various suppliers. In such cases, the 'coachbuilder' who supplied the vehicle to its end-user acted as little more than a co-ordination and marketing organization. In fact, the business could be positively incestuous at times: on occasion, two of the major players might even get together to produce a vehicle.

It is impossible to say for certain how many custom conversions of Range Rovers were carried out; and the position is complicated by the fact that some vehicles were actually returned to the coachbuilders after some years in service for a refit or a major transformation. However, at the height of the boom period in the early 1980s, custom conversions were probably being carried out at a rate of 250–300 a year. At a guess – and it can be no more than that – the total number of vehicles affected must have been between 1,500 and 2,000.

THE SUB-CONTRACTORS

Several of the companies already mentioned in Chapters 7 and 8 acted as sub-contractors to the custom-builders operating in the Middle Eastern market. Thus, Carmichael, Scottorn, SMC Engineering and JNR Motors occasionally supplied three-axle chassis, Overfinch (née Schuler) provided engine conversions and suspension upgrades, and JE Motors provided engines. There were also other companies who took on sub-contract work.

Notable among these companies was A E Smith of Kettering, an old-established firm of coachbuilders whose work had been mainly on PSVs and commercial vehicles. Their involvement in modifying Land Rovers for safari and overland trekking work led them to Range Rover conversion work in the mid-1970s. Their first conversion was to turn a two-door vehicle into a four-door estate on the 136in (3,454mm) wheelbase a type of conversion which they pioneered. Some conversions went out under their own name, but others were done on behalf of the major companies. A E Smith did a lot of work for Wood and Pickett, and did many of their four-door conversions on the standard wheelbase. The company's repertoire included 10in (254mm), and 36in (914mm) chassis extensions; four-door estate bodywork to match; open-back 'troop carrier' vehicles and pick-ups; and convertible Range Rovers with both two and four doors.

A great deal of conversion work was also carried out by Golde of Basingstoke, better known for their sunroofs. They worked on all kinds of conversions, but their particular speciality was convertibles, and they provided these for most of the big names in the business. Grand Prix Metalcraft of London also did some conversion work, both for other companies and under their own name.

THE BIG NAMES

The boom period for Range Rover custom conversions lasted from the end of the 1970s to the end of the 1980s. By then, the Middle Eastern market had reached saturation point, and many customers were demanding newer four-wheel drives from Toyota and Nissan as the basis of their custom conversions. Nevertheless, the Range Rover was still widely respected for its abilities, and small numbers of custom conversions were still being ordered as late as 1994.

Exactly who produced the very first custom conversion will probably never be established for certain, but the most famous name of the period was undoubtedly Wood and Pickett. FLM (Panelcraft), who had pioneered the four-door body conversion, also carried out luxury conversions but were never quite as well known. The end of the 1970s and early 1980s saw four more companies enter the field, although all were short-lived. These were Rapport, Symbol, Chameleon and Elektiar, all primarily marketing operations rather than coachbuilders proper. Also significant in this period were Glenfrome and Townley, but only the former survived into the 1990s. The sole survivor by the mid-1990s was Vantagefield, which had opened for business in 1982.

Wood and Pickett were a London coachbuilding company founded in 1947, and during the 1960s and 1970s they established a reputation as customizers of Minis. They turned to Range Rover work in the mid-1970s, sold out to the Henly Group at the end of the 1980s, and then closed down in the early 1990s when the customized Range Rover market collapsed.

The earliest Wood and Pickett Range Rovers were simple four-door conversions, sometimes with Recaro seats to improve comfort and flared wheelarches to cover wider wheels and tyres. However, the company soon branched out into custom interiors, and by the end of the 1970s had introduced a restyled front end using Rover SD1 headlamps in a sloping front panel. This was commonly known as the Sheer Rover front end, although in fact a majority of Wood and Pickett conversions for the Middle East wore 'Sheer Rover' badging; the name was derived from the Arabic word 'shir', which means 'lion', and Sheer Rover conversions usually had a running lion badge on the sides of their scuttle panels.

The early 1980s saw Wood and Pickett introduce a whole range of other conversion options. Convertibles were known as Goodwood models, and came in two-door form on either the standard or 110in (2,794mm) wheelbase, or in four-door form on the standard wheelbase with a 10in (254mm) addition to the rear overhang. The 109in (2,769mm) wheelbase estate was called a Warwick; Epsom was the name given to the 118in (2,997mm) model; and the 136in (3,454mm) estate was known as the Windsor. The Ascot was a 118in (2,997mm) model with its roofline raised by 4in (102mm), and a similar increase in height was available for 100in (2,540mm) models. A three-axle 6x4 estate with a 136-inch wheelbase was also available in two-door or four-door form, and this was known as a Cheltenham 6. Other models included the Newbury Falcon, a version of the Aintree Multi-Role vehicle (*see* Chapter 7) designed for falconry.

In the mid-1980s, Wood and Pickett announced a vehicle they called the Longchamp, a completely restyled two-door Range Rover which was specifically intended as a rival to the Glenfrome Facet (*see* below). Its styling seemed outrageous at the time, although proved to have anticipated elements of some later Japanese 4x4s, and the Longchamp was produced in a strictly limited edition.

A strikingly individual four-door conversion by Wood and Pickett. There was a detachable hood to make the vehicle weatherproof.

Wood and Pickett were renowned for their high-quality interiors, although the extra fittings in this vehicle clearly stretched their ingenuity to the limit.

This open-top vehicle clearly shows the four-door body conversion characteristic of FLM (Panelcraft).

Wood and Pickett were particularly noted for their luxurious interiors, which often included the Recaro seats for which the company was the sole UK importer. Dralon or leather trim was often employed, in either case available in a wide range of colours to suit the customer's taste. There was also a special facia known as the Margrave (the Mini Margrave had been one of the company's better-known conversions), which incorporated non-standard instruments in a stylish wooden panel, the whole being trimmed in leather. In addition, the company offered a variety of grille designs, additional interior equipment (such as refrigerators and televisions), and cosmetic changes such as chromed bumpers and wheels.

FLM (Panelcraft) developed their Range Rover custom conversions on the back of their successful four-door body conversion. They offered estate conversions with 110in (2,794mm), 118in (2,997mm) and 136in (3,454mm) wheelbases, with the option of six doors on the latter. Convertibles were available as two-doors on the 100-inch wheelbase, and elevating hunting seats were offered with all wheelbase lengths. There was also a three-axle model, based on the Scottorn 6x6 chassis. FLM (Panelcraft) also offered interior retrims and engine conversions to meet customer requirements.

Rapport Automotive were associated with Chris Humberstone Design, a company which had got into the Range Rover market in a small way during 1979 by importing examples of the Australian

FLM (Panelcraft) turned this two-door Range Rover into a convertible, lengthened the rear overhang by 12in (305mm) and restyled both front and back ends.

4.4-litre V8 and offering performance conversions. Their next move was to offer turbocharged Range Rovers (using an Allard turbocharger), and they soon started to supply four-door conversions (known as the Quadraporte), redesigned front ends using Ford Granada light units (the Droop Snoot), and retrimmed interiors with some tastefully redesigned dashboards in wood and leather. A two-door convertible with power-operated top became available under the name of Starlight, and there was also an open-backed Huntsman, with inward facing bench seats in the load bed.

In the early 1980s, Rapport announced the Excelsior, a four-door conversion with a 136in (3,454mm) wheelbase which could be had as either an estate or what Rapport called a 'state landaulet', with a convertible top over the rear seats only. One particularly notable example of the Excelsior had rear doors longer than the fronts, large blind rear quarters with dummy landau irons, and a mock Rolls Royce grille. Three-axle models could also be had with either 6x4 or 6x6 drive configurations, and Rapport continued to offer turbocharged and 4.4-litre engines, together with Borg Warner automatic transmissions and Schuler automatic conversions with Ferguson Formula anti-skid brakes.

Rapport ran into financial difficulties in the early 1980s and closed down, but their operations were transferred to another company within the same group, called Symbol. Their Starlight and Quadraporte conversions (and possibly others) therefore reappeared under the Symbol name. However, Symbol ceased trading in approximately 1984 before introducing any unique new designs.

Trading from an address in London's Park Lane next door to the showrooms used by Rapport and Symbol were Chameleon. This company was already known for its Mercedes-Benz stretch-limousine conversions before entering the Range Rover market. Chameleon offered four-door estates on the 136in (3,454mm) wheelbase (almost certainly built by A.E. Smith), and a variety of three-axle models. These came in either 6x4 or 6x6 versions, and were based on two-door or four-door Range Rovers. Chameleon also offered a model with a stepped roof giving 8in (203mm) of extra headroom, which was not unlike the Carmichael Clansman in overall conception. However, the company did not produce many Range Rover conversions and appears to have stopped marketing these vehicles by the mid-1980s.

The early 1980s also saw Elektiar enter the market, with a single design. The company appears to have been based in Riyadh, Saudi Arabia, and simply to have had a London office in New Bond Street, although there are indications that it may have been associated with Glenfrome (see below). The Elektiar was a 136in (3,454mm) six-door model, with eight forward-facing seats and two pairs of removable inward-facing benches in the rear, so giving a total of twelve seats. A wood-and-leather dashboard incorporating Jaguar instruments was part of the package, the upholstery was in leather or leather with Dralon inserts, and there was a separate air conditioning system for the rear of the vehicle. However, there is no indication that Elektiar ever expanded their model range, and they had probably disappeared from the Range Rover scene by the mid-1980s.

Glenfrome Engineering of Bristol, or Glenfrome of England as they liked to be known, were one of the longest-surviving Range Rover custom converters. They started operations in 1981 and did not close down until the early 1990s. Among their conversions were some particularly imaginative designs, of which the most

Three examples of the extraordinary Glenfrome Facet under construction in the early 1980s. The body was made of GRP over a steel frame.

famous were the 1983 Facet and 1985 Profile, both featuring radically restyled coachwork from the pen of Dennis Adams, stylist of the Marcos sports car.

Glenfrome's conversions were all given type names, but each one could be individualized by cosmetic details such as four-headlamp front ends and special paint schemes. The Ashton was a two-door convertible with power-operated soft top and a 10in (254mm) chassis extension (although early examples were on the standard wheelbase). The 109in (2,769mm) stretch was called the Filton, the 124in (3,150mm) model the Clifton, there was a 133in (3,378mm) model called the Westbury, and the longest stretch was called the Portway and had six doors on a 136in (3,454mm) wheelbase. Notchback models were available on the 124in and 133in wheelbases, in each case known as Dyrham models, and the Glenfrome Six was based on a 6 × 4

Glenfrome's 133in (3,378mm) wheelbase model was called the Westbury and featured rear doors which were longer than the fronts.

One reason for the extra length in the Westbury was to make room for a comprehensive in-car entertainment console, though each vehicle was built to a client's individual specification.

three-axle chassis and could be supplied with a variety of different bodies. These ranged from pick-ups to stepped-roof models generally similar to the Carmichael Clansman.

Glenfrome also built one three-axle model on a 146in (3,708mm) wheelbase, with distinctive wheelarch styling. This was called the Saha Six. It was built for the Malaysian Government and was intended as a hunting vehicle. However, it did not delight its owners and was returned to Glenfrome after a few months for a complete refit, later being sold on through Symbol in London. Finally, Glenfrome also offered open-back vehicles, their models being known as the Falcon (available with

A Glenfrome Dyrham on the 124-inch (3,150mm) wheelbase, complete with the company's trademark eight-lamp front end and chrome grille.

Glenfrome's Saha Six remained a one-off.

The customer who asked Glenfrome to build him this one clearly intended to take enough spare wheels for all eventualities.

either two or four doors) and the Hawk (a two-door fitted with an enclosed cab).

Townley Cross-Country were established Land Rover dealers in Eltham, London, and in the early 1980s were the only Land Rover authorized export dealers in the UK. Their sideline in customized Range Rovers developed naturally from that position, and until 1988 they offered a wide range of customized vehicles to Middle Eastern clients. These included two-door and four-door convertibles, three-axle 6x4 and 6x6 vehicles with landaulet or estate bodywork, six-door estates on the 136in (3,454mm) wheelbase chassis, and the Townley Ranger, an open-back vehicle similar to the Glenfrome Hawk but more highly specified. Townley also offered more limited custom conversion work, including simple retrims and special paintwork on otherwise standard vehicles.

Most exotic of all the Townley conversions were a matched pair of Desert Rangers – the last of their kind – built for a royal customer in the Middle East in 1985. The smaller vehicle came on a 118in (2,997mm) wheelbase, while the larger was

Townley's most exotic creation was the six-wheel Desert Ranger. After the vehicle had been lengthened, widened and fitted with a raised roof and 5.7-litre Chevrolet engine, it bore little resemblance to a standard Range Rover. This one was the last of several similar vehicles built and was completed in 1985.

a six-wheeler on a massive 154in (3,912mm) wheelbase; both were also widened by 6in (152mm). The six-wheeler had inward-facing seats behind the driving compartment, while the four-wheeler had three rows of forward-facing seats. Both had raised rooflines, the four-wheeler with a step in the roof, and both were lavishly equipped and trimmed inside. The larger vehicle, not surprisingly, had a 5.7-litre Chevrolet engine installed to cope with its extra weight.

Longest-lived of all the custom conversion specialists is Vantagefield of London, which opened for business in 1982. The company prides itself on establishing a customer's precise needs (as distinct from humouring his whims) and on delivering a vehicle which is tailor-made to suit. Obviously, no two vehicles the company has built are exactly the same, but there have been a number of basic types, all of which have their own model-names.

Thus, a four-door on the standard wheelbase is known as a Regency when fitted with luxury options, but it becomes a Tiger when fitted with the 4.5-litre Dakar engine supplied by JE Motors (or, more recently, by John Eales Development). With a stepped roofline, such as was provided for one client in order to accommodate a wheelchair-bound passenger, the vehicle

Vantagefield built this six-door vehicle in 1990 as a tour bus for a Middle Eastern client who liked to take guests around an island where he kept exotic fauna and flora.

This mid-1980s Vantagefield conversion was designed as a wheelchair carrier and was fitted with a hydraulic lift at the rear.

This Vantagefield Chamois convertible was a four-door on the standard wheelbase and featured a power-operated roof.

The 109in (2,769mm) wheelbase Vantagefield Rapier conversion was a particularly well-proportioned vehicle. This early 1990s example had elevating hunting seats.

becomes a Ranger. Still on the standard wheelbase, there are fully open and closed-cab/open-back conversions known as Trekkers, and two-door or four-door convertibles bearing the Chamois name. Rarer are the Sceptre models, of which just two had been built by the mid-1990s; these had notchback styling, again on the standard wheelbase.

The next size up in the Vantagefield catalogue is a 109in (2,769mm) wheelbase, and four-door estates on this are known as Rapiers while convertibles are called Starlights. Estates on the 118in (2,997mm) wheelbase are called Westminsters, while the 124in (3,150mm) wheelbase ten-seater model is a Traveller. During the 1980s,

A typical custom interior, in this case on a 1989 Vantagefield Rapier conversion.

Vantagefield also offered the Osprey model on the 124in wheelbase, an estate distinguished by a GRP roof which tapered gently upwards towards the rear.

A similar roof was also fitted to some 135in (3,429mm) wheelbase conversions. The standard 135-inch model is however a six-door estate known as a Gazelle. In addition, Vantagefield offer 6x4 or 6x6 three-axle chassis, marketing stepped-roof estates under the Hunter name and landaulet versions called Falcons.

STATE REVIEW VEHICLES

While custom conversions have always been for the very wealthy, and several have been bought by members of Middle Eastern Royal families, there is one type of custom-built vehicle which has no appeal beyond rulers and royalty. That type is the State Review vehicle.

The original State Review Range Rover was built at Solihull in 1974.

Ever since Solihull built the first Royal Review Land Rovers in the early 1950s, VIPs have seen the value of such vehicles. Not surprisingly, it was Queen Elizabeth II who took delivery of the first State Review vehicle based on a Range Rover. The original idea had come from Rover's Managing Director A.B. Smith during 1971, and the Styling Department had started preliminary sketches in January 1972. The vehicle was actually built in May 1974 on chassis number 35506365B, and entered service in March 1975.

An internal memorandum written three years later by stylist and project co-ordinator Tony Poole explains that:

> Production build was standard apart from deletion of roof panel, tailgate, and side glasses. The only non-standard item was that the vehicle went through the assembly line flying the Union Jack (such is the power of in-plant grapevine!).
>
> Final build was carried out in the Jig Shop at Solihull, the rear chassis member was modified to carry a folding access step and the fuel tank was re-positioned. This enabled the exhaust outlet to be moved to the off-side of the vehicle and away from the rear access door. (These) modifications and all bodywork and interior trim design (were) the responsibility of Rover Styling Department.

The key features of this body were its open rear section with high sides, in which the monarch could stand while being driven in parades. Like the rest of the body, this was constructed in steel instead of the aluminium alloy used on standard vehicles; the steel was less prone to ripples and gave a better finish when the Royal Claret paint was applied. The driving compartment contained a single bench seat, and the rear had four inward-facing leather-upholstered seats and a special compartment to hold the

Royal umbrellas! To prevent the upholstery suffering too badly from inclement British weather, a canvas top could be strapped across the vehicle when the rear quarters were not in use; this was later replaced by two GRP panels, because the straps were found to scratch the bodywork.

A second State Review Range Rover was delivered to Her Majesty on 15 September 1990, and was first used at the review of the Battle of Britain parade at Buckingham Palace. This time, the vehicle was built by Special Vehicle Operations at Solihull, although its design followed that of the earlier vehicle quite closely. This second vehicle was based on a 3.9-litre model and at the time of writing (1995) remains the principal State Review vehicle used by the Queen and HRH Prince Philip. The original Range Rover, known as 'State 1', is of course still in service as well.

A different sort of State Review vehicle was built for the Libyan ruler Colonel Ghadaffi by FLM Panelcraft in the early 1980s. This was essentially a landaulet conversion on a standard-wheelbase two-door model. The centre section of the rear seat folded forwards to give the Colonel somewhere to stand, while behind the seat was a padded box on which he could sit when tired and still be visible to the crowds who had come to see him. This vehicle was finished in green with a black landaulet top and black vinyl roof over the driving compartment.

Glenfrome also built a State Review vehicle for an unidentified (but probably Middle Eastern) client. This was based on a 1984 four-door Range Rover Vogue, converted to two-door configuration with a review platform in the rear. There were fixed rear steps and a raised dais for the reviewer, together with grab rails at the sides and a windshield for the VIP at the rear of the main roof. The vehicle was finished in metallic gold with a red interior and review bay.

For the Pope's visit to Britain in May 1982, Ogle Design were contracted to build

Glenfrome's interpretation of the Review Vehicle theme for a Middle Eastern client was quite different from the British vehicles.

two very special review vehicles, which became known as 'Popemobiles'. These were based on two-door models, and had a specially raised rear section with bullet-proof glass up to shoulder height. The review bay contained bench seats for four people, and there was a fixed rear step with running boards below the doors – both items designed for security officers to stand on. The vehicles were painted white and trimmed in blue, and evidently pleased the Pope considerably. He took one (VDU 100 X) back to the Vatican with him, leaving the other (VDU 111 X) in Britain, where it is now owned by an enthusiast.

Lastly, the RAF Police also ran a State Review Range Rover until 1994, registered as 45AJ49 and based on a two-door model. This had an enclosed cab, side opening rear door, and a detachable perspex roof to cover the review bay in bad weather. While less obviously luxurious than the Royal Review Range Rovers, it was nevertheless a very distinctive vehicle.

The two Popemobiles were built to a unique design by Ogle.

The RAF also had a Review Range Rover, seen here with the Perspex top cover used in poor weather conditions.

9 Individuality: The Range Rover Aftermarket

This chapter will look at the various types of accessories available for the Range Rover, and has been divided into two sections: on-road and off-road. The first section looks at accessories that are more form than function, while the second section examines the vast range of accessories that are genuinely useful if you intend to take your Range Rover seriously off-road.

STREETWISE RANGE ROVERS

You only have to drive the streets of any major city in the UK to see Range Rovers that will never go further off-road than one wheel on the kerb as the owner pops into the local patisserie. Such models are normally late-specification Vogues or Vogue SEs, sporting almost every optional accessory that can be fitted to make the vehicle stand out from a crowd.

Many accessories of this type are available from Land Rover Genuine Parts and can be added by the dealer at the time of purchase. Others are made by a Range Rover aftermarket specialist like Bearmach or Allmakes, and can be purchased via non-franchised dealers or direct from the company itself.

Side Steps

Some of the hardy off-road fraternity like to have a good laugh at Range Rover side steps, reckoning that their fitment is a sure sign that the vehicle will never go off-roading in its life. However, what these people fail to consider is the difficulty that some people have in climbing in or out of a vehicle that is as high as the Range Rover. Although side steps undoubtedly do impair the Range Rover's off-road ability, if they help you or your passengers make a more dignified entry or exit, why should they not be fitted?

Conventional side steps are available from both Land Rover Genuine Parts and other independent accessory specialists, and both types just bolt onto the chassis for a secure mounting. However, there is another type of side step that is even more ingenious. Called the Top Step, this device electronically raises and lowers itself when the door is opened, making access very easy. All the electrical components are placed well up under the body, so only the most serious off-roading is likely to cause damage – a clever solution to the problem of side steps grounding.

Wheels and Tyres

As Chapter 7 shows, most of the high-performance Range Rover specialists recommend fitting Avon Turbospeed tyres for the ultimate in high-speed road-holding. However, the fact that these tyres incidentally look wide and purposeful has not been wasted on owners who wish their Range

Rovers to stand out from a crowd. Fitted with a nice set of Genuine Parts alloy wheels in conjunction with a Range Rover bodykit (from TWR or Rimmer Brothers), low-profile tyres give the vehicle a whole new sporty look; something that was perhaps beyond the wildest dreams of the vehicle's designers.

Earlier models can be made to look different by fitting the classic white eight-spoke wheel, some Compomotive alloys or, indeed, a set of Discovery steel wheels. In all, it is amazing what a completely different set of wheels and tyres will do for the appearance of a vehicle – a fact that has been exploited by a number of specialist wheel and tyre suppliers.

Bull Bars

Bull bars were originally designed as an off-road accessory, intended to protect the front of the vehicle from damage caused by collision with stray animals in the Australian outback. Whether a bull bar actually improves the appearance of a Range Rover or not is a matter of taste; what is undeniable, however, is that a bull bar makes the vehicle look more aggressive. It also makes it look more 'authentic', suggesting that the owner really does spend his (or her) weekends exploring inhospitable regions with the vehicle. There is also an argument that bull bars do more damage to pedestrians than the front of the vehicle would, and at the time of writing there is some controversy over this issue. Like it or not, however, only an outright ban will stop owners fitting bull bars to their Range Rovers.

Available in a variety of shapes, sizes and finishes, bull bars can be bought from virtually every company specializing in 4 × 4 and off-road accessories. Each one differs in its basic design, with some versions (especially the chrome finish type) obviously designed for looks rather than practical protection. Finishes are as diverse as the range of styles, although black powder coat on steel appears to be the most popular. Beware of small nicks in the plastic coat however, as the steel bar will soon rust if not treated quickly.

Leisure Accessories and Long-Distance Travel

As the Range Rover became more and more ingrained with a 'leisure' image, so the accessories for the outdoor types became more diverse. Land Rover Genuine Parts seem to have the main market here and their 1995 parts catalogue for the Range Rover Classic lists no fewer than five types of roof rack, which (according to the catalogue) are suitable for carrying skis, pieces of antique furniture, surf boards and fence posts.

The long-distance traveller can also eschew the conventional open roof rack with a tarpaulin and buy an aerodynamically designed, fully enclosed ABS luggage box that sits at the back of the Range Rover roof. Lockable and featuring gas struts for effortless opening and closing, the luggage box is a practical and useful addition that goes some way to alleviating problems caused by the Range Rover's comparatively small load area. Both Thule and Land Rover Genuine Parts produce luggage boxes for the Range Rover.

Land Rover Genuine Parts

'All kinds of accessories for all kinds of people' is Genuine Parts' proud slogan printed on their 1995 Range Rover Classic catalogue. It is hard not to agree, because Land Rover's own accessories brand markets almost every conceivable accessory the

Roof racks vary from the rugged heavy-duty type seen here …

Range Rover owner might want to have.

From winches to bicycle carriers, mobile telephones to gun boxes, no leisure pursuit in which the Range Rover might find itself is not covered. For horse lovers there is a riding tack rack; dog lovers can prevent man's best friend from getting the seats muddy with a dog guard; baby can be kept safe in a rear-mounted baby seat, while rear seat passengers can be kept amused in the Range Rover LSE with the Genuine Parts cocktail cabinet and fold-away picnic trays.

Other accessories from Genuine Parts include: electric slide and tilt sunroof, sheepskin seat covers, the Land Rover refrigerator, fire extinguishers, a cockpit handlamp, snow chains and a first aid kit.

For those who wish to go beyond accessories for the vehicle, Genuine Parts can also supply a very high quality range of clothing and luggage all with a 'country' theme; while the expensive but elegant crystal model Range Rover must be, as the brochure says, the ultimate gift for the enthusiastic Range Rover owner.

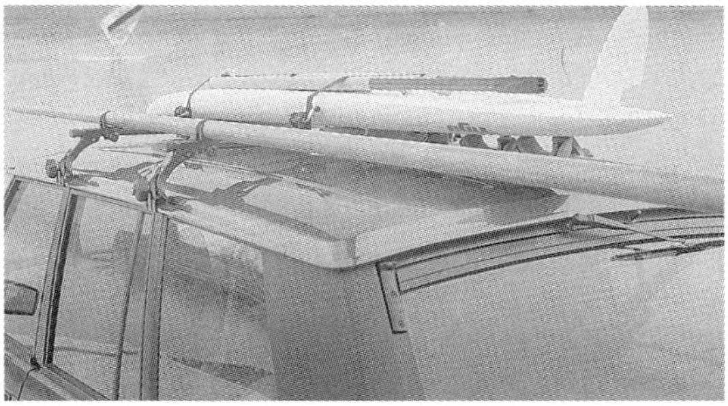

… to the rather less rugged (but no less practical) type in this picture.

PREPARATION FOR OFF-ROADING

The extent to which you prepare your Range Rover for off-road use will naturally be governed by the use you intend to make of the vehicle and by your budget. This section takes a look at what some users have done and highlights some of the accessories and equipment available. There are, of course, many others.

Winches

If you propose to venture off-road alone, a winch is one of the most useful pieces of off-road equipment that you can have, because it will allow quick and efficient self-recovery. Similarly, if your Range Rover is really bogged down in thick mud, a winch can be a safer and gentler way to extract it than the traditional 'snatch' type of recovery using a second vehicle.

Winches have been available for the Range Rover since the early 1970s and have evolved along with it. In the beginning, capstan winches driven from a power take-off (PTO) on the vehicle were the most common variety; but changes in the vehicle's design made different winches necessary and since then both hydraulic and electric types have been used.

The original Range Rover capstan winch was made by Fairey Winches of Tavistock (which later became the UK division of Superwinch) and fitted in the space behind the front grille just in front of the radiator. It was driven by a power take-off on the gearbox and gave smooth but powerful operation on tickover. As the capstan drum was fitted behind the front grille, this had to be equipped with quick release levers to permit full access to the winch itself. The UK Coastguard Service had capstan winches fitted to the small batch of Range Rovers they took in the mid-1970s, although this type of winch was otherwise very rare.

Later additions of an oil cooler and an air-conditioning radiator on the Range Rover meant that there was no space to fit the winch equipment, so the capstan winch was finally phased out in the 1980s. Its replacement was the hydraulic winch, but once again the complexity of the mounting and the necessity of a PTO unit to power the winch meant that such winches remained comparatively rare. Nevertheless, in terms of sheer performance, these winches stand out for their consistently smooth operation no matter how heavy the load. There is little that can faze the hydraulic winch and of course it does not depend, like the later electric winches, on battery power.

By far the most popular type of winch, however, is the electrically driven type. Such winches have been available for the Range Rover for nearly twenty years. It is not entirely clear which model was first fitted to a Range Rover, but the excellent Warn 8274 (which was first announced in 1974) was certainly being used on Range Rovers in the mid-1970s.

Like the capstan winch before it, the 8274 fitted behind the front grille, although its construction ensured that the grille had to be angled outwards at the bottom in order to clear the winch motor. Once again, quick-release levers were added to the grille and the drum could be clearly seen by the operator. However, oil coolers and air conditioning radiators added behind the grille in the early 1980s prohibited the use of the 8274 on later Range Rovers, so instead Warn and other companies offered a variety of low-profile versions that could be fitted behind the bumper.

These winches are wholly compatible with the modern Range Rover, as they sit well out of the way; some models are actually concealed behind the front spoiler. Unfortunately, this position prevents the operator from seeing the winch drum in operation – something that is very important to ensure that the cable feeds in correctly. To get around this, companies such as the Southern Winch Centre and BRB have designed winch bumpers to replace the Range Rover's standard bumper and these site the winch away from the body. Although most of these bumpers do little to improve the Range Rover's looks, they do

The early capstan-type winch is just visible behind the grille of this 1974 Range Rover operated by H. M. Coastguard Rescue Service …

… but electric types have been much more popular in recent years. This one was made by Warn.

offer a clever solution to the problem of winch fitment to late-model Range Rovers.

With electric winches, it is advisable to fit a heavy-duty battery in place of the standard type, because the winch does use a lot of current on particularly arduous pulls. Further back-up can be provided by using a twin-battery system with a split charge facility. If one battery goes flat, you can simply change to the other and recharge the original by running the engine.

Off-Road Tyres

Few things better demonstrate the compromises inherent in the standard Range Rover than its tyres. In simple terms, tyres which run quietly and give good road-holding on tarmac will soon lose traction in mud; while tyres with a tread which is aggressive enough to claw through deep mud will be intolerably noisy and will give poor road-holding on tarmac. The standard tyres fitted by the factory were always reasonably good at fulfilling both functions, but never outstanding at either. It is also interesting to note how the tyres fitted to Range Rovers after the mid-1980s gradually shifted the bias towards road work and away from off-road work; earlier types are much better in the mud and much noisier on the road.

If you intend to take your Range Rover on muddy terrain very often, you will quickly learn to appreciate the benefits that proper off-road tyres can give. Such tyres are available from a variety of different manufacturers and each type has been designed to excel in certain conditions. It is therefore worth bearing in mind how much road work you intend to do before buying a set of tyres that can wear quickly on road.

A favourite tyre size among off-roaders who use Land Rover utilities is the 750 × 16. It is worth recognizing from the outset that these tyres cannot be fitted to a Range Rover unless the front wheelarches (and perhaps also the road springs) are first modified. On an unmodified vehicle, the large 750 tyre is likely to foul the bodywork when the suspension is fully compressed, especially at the front if the steering is on full lock.

However, quite a few off-road enthusiasts have successfully fitted the capable 750 versions of the Firestone SAT tyre, by using higher springs and cutting away the front wheelarch so that the tyre does not come into contact with it. Such a modification should not be undertaken lightly, though, as the cost for two replacement front wings is fairly high if it goes wrong!

Some very good off-road tyres are made by the American company BF Goodrich, and the raised white lettering on some types also gives them an appeal to Range Rover owners who have no intention of getting their tyres muddy. The main aim of the BF Goodrich tyres is in any case a compromise between on-road and off-road ability. By far the most popular of the BF Goodrich tyres with Range Rover enthusiasts must be the Trac Edge; a tyre that is speed rated to over 100mph (160km/h) and also performs excellently off-road. Use of the Trac Edge over a long period of time is quite possible; as the authors' own tests have shown, the tyres last for up to 40,000 miles (64,000km).

Other tyres suitable for the Range Rover include the Mud-Terrain (which has a more aggressive tread pattern than the Trac Edge), the Baja (which is very expensive and more suited to sand than mud) and the All-Terrain (a good all-round tyre, biased more to tarmac than for off-road use).

General Grabbers offer a similar performance to the BF Goodrich Mud Terrain off-road, although some users reckon that the Grabbers have the edge on slippery mud. Either way, the standard General Grabber is an excellent tyre both on and off-road, and the Grabber range includes tyres that are biased for road use and vice versa.

Although Michelin is no stranger to the off-road tyre market (the famous XCL has been used on the Camel Trophy for many years), the French company launched a range of tyres especially for the 4×4 market (the Range Rover included) in 1994 to the acclaim of the international press. The tyres fall into four groups, with which Michelin aims to provide a tyre for every purpose, off-roading included.

Roll Cages

Roll cages are designed to give additional strength to a vehicle's bodywork and to prevent the roof and sides from caving in if the vehicle rolls over. Every Range Rover owner who fits one hopes that it will be a complete waste of money, but if you are contemplating some serious forays beyond the tarmac, a roll cage might be a worthwhile investment to give that extra peace of mind as you attempt a difficult side slope.

Safety Devices manufacture the most well-known type of roll cage for the Range Rover and this UK manufacturer is the exclusive supplier to Land Rover for the Camel Trophy vehicles. The Safety Devices

Range Rover roll cage is fully internal, although you will need to cut a couple of holes in the dashboard for the cage to fit. The SAS are said to have fitted Safety Devices roll cages to their Range Rovers, so this is perhaps as good a recommendation as any!

One disadvantage of the internal roll cage is that it uses up some space in the passenger cabin. For those who worry about this problem, many of the UK's roll cage manufacturers are quite willing to build one-off cages that are half external. These usually have the front supports going down through the scuttle panel (as opposed to through the dashboard); they then fit flush to the windscreen and roof, with the rear support brackets coming down behind the front seats. Some customers have chosen to continue the cage internally to the rear loadspace, although this is not necessary if you only normally carry a front-seat passenger.

The cage supports can either bolt onto the floor, or (for stronger protection) be extended round to the chassis itself. We would recommend the latter. We have also seen some cages that have been fixed to the front scuttle panel and rear floor only. Suffice it to say that this type would offer little protection in the unfortunate event of a roll over!

Underside Protection

If you are considering taking your Range Rover off-road, the first accessory you should buy is some type of front steering arm protector. Steering arms are all too vulnerable and it is quite easy to bend them on rocks or even compacted mud if you are stuck deep in ruts.

In the UK, the most popular type of front guard is made by Southdown Engineering in Devon, although this is fairly heavy as it is made from steel. The French prefer using more lightweight aluminium guards, and these are manufactured by a variety of off-road specialists, most notably ETT (*see* Chapter 11) and Allmakes. Aluminium guards cannot be purchased in the UK, so if you want one of these you will have to go abroad.

The next area of protection that should be considered is the axle differentials. Once again, Southdown Engineering produce an excellent guard plate for the front differential and axle and this has the added benefit of protecting the track rods as well.

An underbody protection plate to protect the vulnerable steering arm is a sensible addition to a Range Rover used for serious off-road work.

The French again offer a different solution. Called the *'coquille'* (meaning 'shell'), it is a semi-circular steel cup which is welded over the differential housing itself, thus providing another layer of steel. Used in conjunction with two axle housings for each side of the differential, this offers useful protection with minimal effect on ground clearance.

Apart from the axles and steering rods, the rest of the Range Rover's mechanicals are well hidden above the base of the chassis, thus making damage unlikely. However, the rear overhang and exhaust pipe can ground very easily, often causing quite serious damage. In the UK, there is no type of protection available to alleviate this problem, but once again the French have come up with a solution. In that country, where late-model Range Rovers are used extensively off-road, several different types of rear overhang protection and exhaust mounts are available.

To protect the rear overhang, steel tube is bent into shape to run the length of the overhang and this is fitted to a specially-made rear bumper. On the other end, it bends round and is welded to the chassis, thus making a very strong guard that is able to take the knocks. As for the exhaust, this is rerouted upwards away from the bottom of the bumper, so that the tailpipe exits from a hole cut in the rear bodywork itself. Such a modification is extremely effective, especially as the rear overhang protector shields the tailpipe as well as the bodywork. It really is very strange that these modifications are not also commonplace in the UK.

Snorkels

Even if you waterproof your Range Rover's electrics, wading through deep water is always a risk because of the danger that water might be sucked up into the engine through the front air intake. Although Solihull have mounted the standard air intake as far out of harm's way as possible (on top of the engine on carburettor V8s and above the front wheelarch on others), entering deep water with an unmodified vehicle still remains a risk.

To eliminate this problem, the raised air intake (or snorkel) was developed by Land Rover engineers, although it is now manufactured and sold by the British company Mantec. The snorkel fits a pipe over the standard air intake and diverts it along the inner wing, out through the scuttle panel and up along the side of the windscreen. To prevent rainwater and mud entering the unit, there is a special mushroom-shaped hood that sits on the top. The Mantec snorkel can be used with the TDi, VM Turbo Diesel and 3.5-litre V8s, and there is a version for the 3.9-litre V8 currently under development. Using a snorkel is very effective and engines so equipped continue to run even when they are completely submerged.

Locking axle differentials

With the standard centre differential lock fitted to the Range Rover, drive is shared equally between the front and rear axles. However, as the axle differentials only transfer power to the wheel that can spin most freely, traction can easily be lost when two wheels spin uselessly in mud.

Land Rover have gone some way to solving this problem by fitting Electronic Traction Control to top-model Range Rovers since 1992, although for the ultimate in off-road traction you should fit lockable axle differentials. Two versions are currently available on the UK market: the well-known ARB Air Locker and another air-operated type made by KAM Differentials.

Both versions are fairly expensive to fit, at up to £1,000 per axle for the ARB version at 1995 prices. The KAM unit is slightly cheaper, although this is a relative newcomer to the market. However, as the unit uses the same Bosch compressor that the factory fit for their air suspension, it should prove to be reliable.

Using push-button switches that are fitted inside the vehicle, the locking differential can be simply engaged and disengaged. Engaging one provides an equal power split to both wheels, thus allowing the wheel with more traction to push the vehicle through. However, it does not take a genius to realize that in a situation where no wheel has grip, the system offers no advantages over conventional differentials!

It is very important that the axle differential lock is disengaged as soon as the tricky area has been passed, as strain can be put on both the differential and half-shafts when cornering with the locker engaged.

General Accessories and Additions

As well as the major accessories needed for off-roading, there are several smaller bits and pieces that we recommend for off-road Range Rovers. They are available from a variety of suppliers, including Land Rover Genuine Parts, Bearmach and Allmakes.

To protect the Range Rover's interior from the effects of a serious off-road session, fit waterproof seat covers and rubber floor mats. If you use your vehicle on-road as well as off-road this is particularly important, as you do not want to get into the vehicle on a Monday morning with your best clothes on, only to remember too late that the seats and carpets are covered in mud from your weekend off-roading!

Front and rear light guards offer protection for expensive light lenses, shielding them from rogue branches and other off-road hazards. However, when fitting them, make sure that they are completely

Light guards are a popular accessory, although they are often bought for cosmetic purposes only.

secure, as rear ones are particularly susceptible to being ripped off during serious off-road excursions, thus damaging the bodywork.

On 1987-model Range Rovers, the addition of front and rear bumper end caps was one of the most effective parts of the seventeen-year-old Range Rover's tasteful facelift. However, these end caps are notoriously susceptible to being torn away off-road; very early examples had a mounting which tended to twist the bodywork as well when the end cap came away, so Land Rover swiftly modified that. Of course, if you do not take your Range Rover off-road that often, replacing an end cap once a year is not too much of a financial outlay; but if you have to do it five or six times a year, it can become an expensive nuisance.

To overcome this, try fitting the rubber bumper end caps that became available on 1983 models. These do not wrap around the bodywork like the post-1987 versions, so cannot be knocked off easily. The caps are available as a Land Rover Genuine Part.

On Range Rovers that are used regularly off-road, higher and/or heavier duty springs and shock absorbers along with a set of polyurethane Polybushes can produce a firmer off-road ride, as well as being more tolerant of severe off-road punishment. For the ultimate off-road suspension system, try fitting double shock absorbers and heavy duty springs. One of the most popular spring combinations for off-road Range Rovers is diesel-specification springs at the front, with Genuine Parts Heavy Duty coil springs fitted to the rear. These raise the body slightly and give a firmer, more controlled ride in the rough.

Off-Road Bull Bars

As mentioned earlier, the majority of Range Rover bull bars are of little use off-road. They merely get in the way and can even increase the amount of damage caused when you hit something, as the bull bar folds back into the bodywork.

By far the best bull bars to use for an off-road Range Rover are the ARB and BRB

Replacement bumpers by ARB allow a high-lift jack to be used. This picture shows an ARB rear bumper in place.

types, which have been specifically designed with off-roading in mind. Some versions come with a winch mount, although this is not compulsory. Both the ARB and BRB versions incorporate eye-holes for a hi-lift jack, and as a hi-lift cannot be used on a Range Rover without such a bull bar, serious off-roaders should consider them very carefully. The Australian company ARB also make a rear bumper that matches their front bull bar, and once again this is particularly useful if you want to use a hi-lift jack with your Range Rover.

Useful Equipment

As well as accessories bolted to the vehicle, every off-roader should have the minimum amount of off-road equipment to help when the vehicle becomes stuck. For the Range Rover, our suggestion would be: a Kinetic tow-rope, a selection of both 'D' and bow shackles (painted bright yellow for easy identification), a tree strop and a shovel. In addition, we recommend investing in a New Concept Airbag or hi-lift jack. The air bag (essentially a 'balloon' strong enough to lift the Range Rover, which is inflated from the vehicle's exhaust) is a good alternative to the hi-lift, especially for vehicles without the eyeholes which this type of jack needs.

Other useful pieces of equipment include the Tirfor hand winch (which needs only muscle power to extract a Range Rover from the most glutinous mud imaginable) and the suction-fit Multi-Light. The latter attaches onto any flat part of the Range Rover's body by means of a powerful rubber sucker and gives instant light using the Range Rover's cigarette lighter socket. It is particularly useful if you find yourself off-roading or (more likely) stuck in the dark.

Equipping the Range Rover for Expeditions

Although not perhaps the most obvious choice for an overland expedition (the Defender 110 or Series III 109 Safari remain ever popular), the Range Rover has much to commend itself for long distance expeditions. Permanent four-wheel drive means that the vehicle can cope with almost any type of terrain; the true on-road comfort of the Range Rover can prove invaluable for any fast on-road sections of the expedition; and the vehicle's comfortable interior is streets ahead of the Defender – an important issue if you intend to be living in your car for weeks at a time. So how do you prepare a Range Rover for a long-distance off-road expedition?

Firstly, you should look at the list of off-road accessories above and assess which of them are necessities and which of them are luxuries. When planning an off-road expedition, you should keep the vehicle's laden weight to an absolute minimum, so you must ask yourself whether items such as a winch or roll cage are unnecessary extra baggage. If the area you are going to has few petrol stations, you will need to consider fitting a long-range fuel tank or carrying extra fuel in jerrycans. Most countries prohibit the carrying of jerrycans on the body, so you will need to think about stowing them in the load compartment or on a roof rack.

The roof racks offered by Land Rover Genuine Parts are unsuitable for carrying heavy loads (they are designed for luggage, skis or surf boards), so you will need to invest in a heavy-duty roof rack made by Brownchurch or Bearmach. Roof racks are not only useful for carrying jerrycans, they can also be used for stowing an extra spare wheel and tyre; even a hi-lift jack. It is also advisable to carry a reasonable amount of spares for your journey, although this

Well-equipped with accessories for off-road use, this Range Rover sports an electric winch and a snorkel-type engine breather.

should be in proportion to firstly the possibility of damage on the trip and secondly the availability of parts in the country you plan to go to.

As a general guide, fanbelts, half-shafts, a complete set of spark plugs, oil and air filters, a distributor cap, rotor arm, universal joints and a track rod would be useful things to have in an emergency.

Before setting off, have your vehicle thoroughly serviced and checked over by a professional garage; if possible a Range Rover specialist. This will ensure that the vehicle is in tip-top condition before setting off. Make sure you have any worn parts replaced before leaving the country. As with any venture, prior preparation and planning should ensure that your expedition goes smoothly, and ideally you should be almost ready to leave a week before your actual departure date. Such preparation will mean that any possible mishap can be dealt with effectively, thus making your trip as enjoyable as possible.

10 Driving the Range Rover

Put a 1970 Range Rover next to a showroom-fresh Classic and, apart from the addition of couple of doors, it is simply astonishing how similar the two vehicles look. However, such similarities (as the preceding chapters have shown) are only skin deep. Improved suspension, engines, gearboxes, brakes and, of course, better soundproofing have nothing short of transformed a vehicle that was always designed to be as comfortable on-road as it was capable off it.

Although twenty-five years have changed the Range Rover little cosmetically, it is from the driver's seat that you truly appreciate the myriad of improvements that the engineers at Solihull have made.

A full page advertisement in *The Times* on 16 June 1970 (the Range Rover's public launch date) bore the proud claim that: 'There are four basic types of car in the world. And you can now own all of them'. This teaser went on to ask: 'What would you call a car that: (a) offered the comfort of a luxury saloon; (b) handled like a high performance car; (c) had the capacity of a high performance car; (d) was as rugged as a Land Rover?' As if readers had not guessed by then, the advert ended with the boast: '[Well], we've made it. And called it the Range Rover. For the first time you can have four cars under one roof.'

Looking beyond the marketing rhetoric, there can be no doubt that the new Range Rover was unlike any other vehicle in 1970; a statement backed up by *The Times* motoring correspondent's article in the same issue: 'The Range Rover is a remarkable

The early Range Rover was a remarkable road vehicle, although it had a number of faults.

vehicle. As a complete all-rounder [it] can commute from suburb to city or cross the Sahara; pull a two-horse trailer, through thick mud if need be, or take the children to school.'

Driving a Range Rover, either on or off the road, is a very special experience – there is simply nothing like it. The vehicle's commanding position puts you literally head and shoulders above other motorists, which perhaps explains the inescapable feeling of superiority most Range Rover drivers experience, even if their attitudes are otherwise politically correct!

Although a 1970 four-speed manual is worlds apart from a 1994 Vogue LSE automatic, the same basic Range Rover qualities remain: excellent on-road ability, high driving position and off-road supremacy.

Most first-time drivers are disappointed with the performance of the standard Range Rover. People tend to think that the large V8 will make it a very fast car, but they forget the other half of the equation; that is to say, low gearing and heavy weight. Although a 100mph (160km/h) off-roader was an impressive novelty in 1970, the Japanese competition were soon matching and improving on the Range Rover's acceleration and performance figures, and Solihull only really began to fight back with the launch of the fuel-injected engines in 1985.

However, driving a Range Rover still remains a joy that has to be experienced to believed, and this chapter aims to help all Range Rover owners to get the best from their vehicles, whether on or off the tarmac.

ON ROAD

The Early Days

As the first chapter in this book explains, the designers of the 100-inch Station Wagon were not initially sure who would buy their new baby. Although the still-sketchy 'leisure vehicle' concept had been identified by Graham Bannock, the inevitably cautious Rover marketing people initially identified wealthy landowners and farmers as potential buyers, assuming that these folk would use the Range Rover as a workhorse during the day and (after quickly washing it out with a hose) step into it for a bit of socializing in the evening.

Accordingly, early Range Rovers were every inch an off-road/on-road compromise. The big V8 engine offered enough power to keep up with contemporary saloons, while long-travel coil springs and a central differential lock ensured a comfortable and sure off-road capability.

However, as the Range Rover began to find its market, complaints started to emerge about the vehicle's road manners. The LT95 four-speed gearbox can only be described as agricultural, with perhaps the largest gaps between gear notches in automotive history! The quiet burble of the V8 was let down by the transmission whine from the main gears and transfer box, and the ratios were comparatively short. So much so, that some drivers of these early vehicles have always preferred to pull away in the 'longer' second gear rather than use first.

Although early Range Rovers are capable of speeds well in excess of the speed limit, few owners pushed them hard when cruising. Transmission and wind noise made long-distance motorway work tiring, and the basic all-vinyl interiors were hardly inviting for an afternoon's hard driving. Early Range Rovers were designed primarily for pottering about on the estate and in this respect they fulfilled their role excellently.

Another on-road characteristic that

customers initially found alarming was the Range Rover's tendency to wallow when cornering. Although the vehicle can actually be pushed much further than most owners would dare, alarming body movement proved a little disturbing for some. Like drivers of the Citroën 2CV, Range Rover drivers simply got used to excessive body roll – although for owners of contemporary saloons it was certainly novel.

Body roll was a problem on Range Rovers until the introduction of anti-roll bars in 1990. Tired springs invariably make the roll worse.

In short, on-road the early Range Rover was more Land Rover than car, although there could be no doubt that it was more tolerable over long distances than the contemporary utilities which bore that name. As the needs of owners shifted from an off-road car with an on-road ability to the other way around, Solihull was keen to supply. During the early 1980s five-speed and automatic models removed many of the criticisms of the early Range Rover.

Changed compression ratios and camshaft profiles also improved the trusty V8's performance, most notably by boosting low and mid-range torque to give quicker acceleration. But it was not until the launch of the electronic fuel injection system that performance-conscious owners were to really get the acceleration they wanted.

Best Ever

By the late 1980s it was becoming clear that few new Range Rovers were being used off the road to the extent that their designers had originally expected. Up until then, Solihull had always shied away from limiting the Range Rover's legendary axle articulation (even if that meant comparatively archaic handling), preferring not to hinder the vehicle's capabilities off-road.

In 1990, however, Land Rover announced the special edition Range Rover CSK, which offered anti-roll bars as a factory fitment for the first time. Undoubtedly Land Rover were testing the water for these suspension modifications, which obviously provoked the desired response, ultimately resulting in anti-roll bars as standard fitment for top of the range models from 1991 onwards.

Further salvation for the road driver came with the bored-out 3.9 and 4.2 litre engines which offered more power and more torque (although the high-revving 3.9 could sound rather busy at high speeds), along with air suspension and the sophisticated anti-lock braking system. In short, these modifications were necessary if the Range Rover was to compete with its new-found competitors, namely top of the range saloons from Mercedes, Jaguar and BMW, and the changes were universally welcomed.

The announcement of air suspension was

by far the most significant change for the Range Rover on-road driver. The four air-inflated rubber bags made on-road travel in the Range Rover even more comfortable, with the vehicle automatically lowering itself at speeds in excess of 50mph (80km/h). Such a move may appear cursory, but as the Range Rover has an unmentionable drag factor, every little helps!

In summary, over a quarter of a century has changed the Range Rover from dual-purpose farm hack to executive cruiser, although the off-road capability has not been compromised. As the demands (and wallets) of the customers changed, so Land Rover altered the Range Rover to suit, producing a vehicle that both fulfilled and exceeded the promises of the original on-road design brief.

Getting the Best from the Vehicle On-Road

So how do you get the best from the Range Rover on-road? Stepping up into the cockpit, first-time drivers soon realize that the Range Rover is a vehicle unlike any other. A commanding view of the road and excellent all-round visibility are standard equipment with every Range Rover, and once you have lived with one for a while, the view afforded by other cars is never the same!

Such a superior driving position makes the Range Rover a very safe car to drive, with the driver being able to literally look over other traffic to anticipate the road ahead. Indeed, this is one reason why the police use Range Rovers for motorway work, many officers preferring the high vantage point of the Range Rover to the out and out high performance of other pursuit vehicles.

However, such height has its disadvantages. The Range Rover's higher centre of gravity inevitably ensures less stability than ordinary two-wheel drive saloons, so driving style has to be altered accordingly. Hard and fast cornering is not

The later Range Rovers were much more reassuring to drive on the road at speed. This one was in fact Turbo Technics' turbocharged 3.9-litre demonstrator, which was being enthusiastically tested by one of the authors when the picture was taken on a clear stretch of dual carriageway.

recommended, although the driver's nerve will have gone long before there is any danger of the vehicle tipping over. Anti-roll bars fitted to later models give much higher levels of stability, although you should never overestimate the capabilities of the tyres, especially in the wet.

With such a tall vehicle, road manoeuvres should be anticipated well in advance; especially at high motorway speeds. Sudden braking or tight turns that would be acceptable in a normal saloon should be avoided wherever possible, especially in early models not fitted with anti-roll bars. If you do feel a loss of control (especially when cornering), back off the power gently and avoid stabbing the brakes suddenly. Hitting the brakes in non-ABS cars is likely to make a situation many times worse than simply backing off on the accelerator.

At the time of writing (1995) there has been much debate in the media about the safety of Range Rovers, most notably after a series of much-publicized accidents involving all makes of 4 × 4. It is inevitable that the tall and heavy Range Rover is going to be more difficult to handle than smaller cars in dangerous situations, but the fact still remains that the Range Rover is probably safer than the majority of cars; not least thanks to its large size and solid steel framework.

With the arrival of front side impact bars (rear also on the LSE) and air bags on the most recent models, there can be no doubt that a late-model Range Rover is one of the safest vehicles on the road today, providing the driver respects its limitations and does not try to drive it like an ordinary car.

OFF-ROAD

The Range Rover is, and always will be, a superb off-roader. Long-travel coil springs coupled to live axles ensure axle articulation which is almost legendary in off-roading circles, while the powerful V8 and turbodiesel engines have enough torque to pull the vehicle through all but the most extreme of off-road obstacles. Part of

Aftermarket handling kits can improve the Range Rover's handling considerably. The cornering speed of this vehicle was high enough to get a standard Range Rover into serious difficulties, but suspension modifications by the French ETT company saved the day.

Chapter 9 outlines how to improve the basic vehicle's off-road abilities still further through the fitment of aftermarket accessories, although the standard vehicle still remains eminently capable.

It is beyond the scope of this chapter to give a thorough guide to off-roading (for further details see books such as *The Land Rover Experience* by Tom Sheppard), so instead we shall look at the main techniques necessary for successful and safe off-road driving in the Range Rover.

Before Setting Off

Unlike straight-line motorway driving, the majority of off-roading situations require a good deal of arm and pedal movement, so it is important that you make yourself comfortable in the driving seat. Adjust the seat so that your arms are slightly bent when holding the steering wheel and make sure that the seat back is in the most comfortable position. The gear-lever and pedals should be within easy reach and all major controls should similarly fall easily to hand.

Excellent axle articulation allows a Range Rover to keep all four wheels in contact with the ground for maximum traction when driving off-road.

Well-equipped for extremes of recreational off-roading, this Range Rover demonstrates what few owners would even dream of attempting. It did *emerge from the muddy pool, intact!*

Regarding the use of seat belts, there are several schools of thought, although the choice ultimately rests with the driver providing it is legal to do so. Wearing a seat belt means greater safety should an accident occur, so we recommend buckling up. Likewise, you should ensure that all passengers are securely fastened in and that they can reach the various grab handles in the vehicle.

When travelling over rough terrain, it is likely that items inside the vehicle will become dislodged with extremes of movement. Therefore, all small items like tapes, pens, de-icer, etc should be stowed in the central cubby box (if fitted) to stop them flying around. Items in the boot should also be secured, with the loadspace cover in place over the top to prevent them from entering the passenger compartment. A metal dog guard placed behind the rear seats gives added peace of mind should the worst happen.

As far as basic vehicle preparation goes, you should make sure that all tyres are inflated to the correct pressure and that the oil and water levels are correct. If you intend to ford a lot of deep water, you should fit the gearbox wading plug and (in the case of the V8) ensure that the distributor and HT leads have been 'waterproofed' by silicon sealant and spray respectively.

As with all Land Rover products, before finally setting out off-road, it is imperative that you engage the low ratio in the transfer box and the centre differential lock (although on late vehicles the centre differential is controlled automatically). Low ratio can be engaged by depressing the clutch and pushing the short gear lever through Neutral into Low. If it is difficult to engage, select second gear and gently ride the clutch while pushing the transfer gear-lever into position; it should then feed itself in easily.

On vehicles with a four-speed LT95 gearbox, the differential lock is engaged by pulling up the round knob on the transmission tunnel. In all other vehicles fitted with a manual centre locking differential, the transfer lever should be pushed to the left to engage the differential lock. An orange warning light will illuminate on the dashboard to confirm that it is working.

With automatics, low ratio should be engaged with the main gear lever in neutral, while the differential lock is operated as above. Post-1990 models, both manuals and automatics, have a viscous coupling that engages the differential lock when necessary, thus relieving the driver of the need to judge the appropriate moment.

Once back on tarmac, the differential lock should be disengaged to prevent potentially damaging transmission wind-up, which can ultimately lead to broken halfshafts and other mechanical components. You will often find that the differential lock does not disengage immediately, and if this is the case, simply reverse for a few metres until the warning light goes off.

Mud and Ruts

As the Range Rover is fitted with permanent four-wheel drive, muddy conditions can be handled with ease. However, with the standard tyres, more power is often necessary in order not to become stuck. Nevertheless, no matter what tyres your Range Rover has fitted, avoid using too much power, as this will cause excessive wheelspin and potential loss of control.

All versions of the V8 engine have plenty of torque, so it is normal to drive off-road using just second or third gear low. You should never pull away in first low as the ratio is far too short; instead choose second or third according to the conditions. The various turbodiesel engines fitted to the

It is vitally important to lock the centre differential (except where a VCU is fitted) when driving in rough terrain. Otherwise, this can happen: all the power is diverted to the spinning front wheels and the vehicle simply goes nowhere.

Range Rover at the factory over the years have been designed or adapted with off-roading in mind, so a similar gear choice applies – there will always be enough low-speed torque to cope.

When travelling in ruts, remember at all times to keep the vehicle moving, especially if the ruts are deep and there is a danger that the vehicle will ground on its chassis. Off-roading guru David Bowyer has coined a useful phrase to describe the best approach to speed – 'Drive as slowly as possible and as fast as necessary' – and this approach should be adopted at all times. Excessive speed may result in loss of control, while too little momentum could mean that you become stuck.

Where possible, let the front wheels follow the ruts, taking special care not to have the wheels fighting against the vehicle's direction. Once a wheel is in a rut it is very difficult to get it back out, so the basic advice is to keep the vehicle moving. If you think that ruts may be too deep for your tyre size, avoid falling in them – instead straddle them, taking care not to slip in!

Wading

Land Rover say that the maximum advisable wading depth for the Range Rover is half a metre (about 19½in, or a thumb's length below the top of the wheel rim). In fact, this can be exceeded with a little care. Providing the engine's electrics (on petrol types) have been waterproofed with silicon grease, the engine is unlikely to cut out – although care should be taken with the engine air intake.

After ascertaining that the water you wish to wade through is not too deep for your vehicle and that the base is firm (this can be checked by walking ahead in waders or by using a stick as a measure), enter the water carefully and slowly. Once the rear axle is in the water, apply the power gently (normally in second gear low) so that the vehicle pushes a small bow wave in front of the front bumper. Keep the power on smoothly, and avoid accelerating too much. Going too fast will cause you to overtake the bow wave, or will throw water over the bonnet and into the engine compartment

Wading through water this deep is no problem for a Range Rover, provided that special attention has been paid to the door sealing rubbers. However, it goes without saying that drivers should never *attempt to drive through water like this without first checking the crossing on foot.*

with possibly disastrous consequences.

Range Rovers cope with deep water fording without many problems, although one consistent fault has proven to be water leaking in through the bottom of the doors. This is due to the construction of the rubber door seals, which unfortunately feature a break at the bottom, thus ensuring that water enters all too easily. This break can easily be sealed using bathroom-type sealant and this is recommended if you intend to do a lot of wading in your Range Rover.

Ascents (Including the Failed Hill Climb)

When climbing hills off-road, gear choice is perhaps the most important key to success. However, the choice of ratio depends very much on the surface of the hill and the type of engine fitted. Below, we shall try and offer some guidelines for most types of vehicles.

Generally speaking, the best advice for hill climbing is to select the highest gear possible, thus preventing loss of momentum and excessive wheelspin which may result in the vehicle coming to a halt. With all versions of the V8 petrol engine, the best gear to use is either second or third low, depending on the length or severity of the climb. Experience will teach you which ratio is the best for each situation, although the authors would recommend using third whenever possible, thus minimizing wheelspin and potential strain on mechanical components. Diesels, on the other hand, seem to climb better using second gear low; although third can be used on shorter climbs. Never use first gear for ascents – the ratio is just too low and you are unlikely to be successful in reaching the top.

If your attempts do prove to be unsuccessful, you will need to employ the failed hill climb technique, which is one of the most important drills an off-roader should know. With the vehicle at a halt on the hill, the top priority is to get back down to the bottom as safely and under as much driver control as possible.

If you feel the vehicle start to slow up on a hill, bring it to a halt by depressing the footbrake and disengaging the clutch. The vehicle should stay stationary on the slope.

Once you are sure that your front wheels are straight, place your right hand at the centre of the steering wheel (in the case of right-hand drive vehicles), select reverse and bring up the clutch. At the last minute let go of the footbrake, and the vehicle should gently drive back down the slope. If the slope is both steep and slippery, you may find yourself sliding backwards with the footbrake depressed. In this case, select reverse as quickly as possible and let engine braking do the work. You can now attempt the hill again using a different gear!

Descents

When descending, you should always use first gear low, no matter how steep the hill. This gives you the maximum engine braking although you should be aware of the inherent risks of using first low, when the vehicle's downward momentum can sometimes exceed the traction available and cause a slide. There is no need to use the footbrake: in fact, depressing the brake pedal while descending may result in all four wheels locking up, which may then see the car slipping down hopelessly out of control, usually faster than it would in second gear low.

The amount of engine braking available is directly related to the compression ratio of the engine. All other things being equal, therefore, the higher-compression engines will give more engine braking; the turbodiesel engines have much higher compression ratios than the petrol V8s and are particularly good in this department. However, even the lowest of the petrol V8 compressions (8.13:1) gives first-class engine braking, which is greatly assisted by the low gearing of the transfer box.

Automatics require a very different approach to descents, and the techniques necessary are covered later in this chapter.

General Good Practice

Off-roading requires a great deal more care than on-road driving, and several basic rules apply for all off-road situations. Firstly, keep your thumbs out of the steering wheel rim, as kickback from the wheel has been known to break the bones of the unwary.

Secondly, keep your feet away from the clutch unless, of course, you wish to change gear. Riding the clutch not only causes wear on the components, it can also result in a loss of control should you engage the clutch accidentally.

Thirdly, read the route ahead. When driving off-road, it is worth anticipating what obstacles or 'dangers' are approaching (as you would on the road), so that you can alter your driving style accordingly. For example, you should build up momentum before you encounter a mud run, and reading the road ahead can also allow you to brace yourself for any shocks should you get it wrong!

Lastly when driving off-road, use the engine and the correct gear to control the vehicle's momentum. Avoid making any sudden speed changes and only use the brakes gently when you want to bring the car to a complete halt. All engines fitted to the Range Rover have been tuned with off-roading in mind, so they are incredibly flexible in their delivery of torque. For example, you can crawl along in third low, but the same gear can also be used to deliver acceleration smoothly when it is required.

Like so many things in life, there is no substitute for experience off-road and the Range Rover is a fine vehicle in which to learn. Its incredible off-road ability is forgiving for the novice, but the vehicle is also accomplished enough to offer something for the more experienced.

This kind of terrain presents no major difficulties for a Range Rover, but drivers should always respect the countryside and should not hack vegetation out of the way willy-nilly in order to get their vehicles through.

Off-road, the driver's nerve usually fails before the Range Rover does. With four up, this example (modified only by the prudent addition of a winch) demonstrates its side slope ability on an off-road course used by Land Rover dealers Dutton Forshaw of Maidstone.

Rules for Automatics

Many experienced off-road drivers have very fixed views on the merits of the automatic gearbox off-road: in a nutshell, they either love it or hate it. To complicate matters further, those who love it are divided on the subject of how it should be used. Is it better to leave it constantly in Drive (descending excepted), or should you use the four available ratios as you would in a manual? In fact, the authors disagree with one another on this point!

We can therefore do no better than to suggest that it is purely a question of personal choice and that there is no 'right' or 'wrong' way of doing things. Whichever way you choose, there can be no doubt that driving the automatic Range Rover off-road is immensely pleasurable, not to say relaxing and easy!

It is on long descents that the automatic gearbox's one failing off-road becomes apparent: engine braking is not as good as with a manual gearbox. On short descents, this is not normally a problem; you simply end up going down faster than you would in a manual. However, with particularly steep or long descents, this is potentially dangerous, as the car could career downhill without the driver being in full control.

In this case, the automatic Range Rover owner would be advised to try the technique of power braking: holding down the brake pedal with the left foot while accelerating with the right. With a little practice, the correct amount of control can be realized, and the Range Rover reaches the bottom of the hill safely. Alternatively, you could try experimenting with the ABS braking system fitted to late Range Rovers while coming downhill, as this is designed to work off-road as well as on the road. Nevertheless, even ABS cannot overcome a lack of traction, and it is advisable to rely on the gears rather than on ABS for long descents.

Air Suspension and Electronic Traction Control

In 1992 electronic traction control (ETC) and air suspension on top-model Range Rovers arrived, and these two features deserve special consideration. Both systems have their value in off-road driving as well as on the road.

ETC works on the rear axle only and essentially uses the brakes to stop a wheel spinning, thus transferring power to the opposite wheel, which in theory should have more grip. On the road, the ETC system is designed to allow the vehicle to pull away smoothly if one rear wheel is on a loose surface (such as gravel) while the other is on firm ground. Off the road, it has an effect similar to that of an axle differential lock (an accessory favoured by some off-road drivers) and allows the vehicle to move when only one of its four wheels has traction. When working, ETC makes a characteristic chattering noise and operates a warning light on the dashboard.

The arrival of air suspension on the Range Rover was hailed as a technological breakthrough and with reason. Like earlier air suspensions, the Range Rover's system offered a smooth and silent ride on the road, but it also provided significant off-road benefits which had never been seen before. The same height-adjustment system which lowered the vehicle to improve its drag coefficient at speed on the road was employed to raise the body to prevent grounding when off-road.

With air suspension, the Range Rover continually monitors its angle over ground, adjusting itself automatically so that the body remains as level as possible. None of the traditional coil spring suppleness is lost and, in extreme off-road conditions, the system automatically pushes a spinning airborne wheel downwards until it regains ground contact and traction. This is, of course, particularly valuable in the Vogue LSE and County LWB models, whose long wheelbases make them more prone to grounding than the standard-wheelbase types.

Typical Road Performance Figures: a Comparison

	2-door 3.5 V8 carb (4-speed)	*4-door 3.5 V8 injection (auto)*	*4-door 3.9 V8 (auto)*	*LSE/LWB 4.2 V8*
0–60mph	14 sec	12 sec	11 sec	11 sec
30–50mph	9.5 sec	4.9 sec	5 sec	5 sec
50–70mph	10.5 sec	7.5 sec	7.3 sec	6.5 sec
Max speed	95mph (153km/h)	102mph (164km/h)	110mph (177km/h)	110mph (177km/h)
Fuel use	15mpg (19l/100km)	17mpg (17l/100km)	18mpg (16l/100km)	18mpg (16l/100km)

	4-door 2.4 VM Turbo D	*4-door 2.5 VM Turbo D*	*4-door 200Tdi*
0–60mph	16.5 sec	15.8 sec	16.8 sec
30–50mph	13 sec	10 sec	8.5 sec
50–70mph	12 sec	11.5 sec	9 sec
Max speed	92mph (148km/h)	95mph (153km/h)	94mph (151km/h)
Fuel use	24mpg (12l/100km)	24mpg (12l/100km)	28mpg (10l/100km)

Off-Road Data

	Standard wheelbase	*LSE/LWB*
Minimum ground clearance	7.5in (190mm)	7.5in (190mm)
Approach angle	35 deg	35 deg
Extended profile, ECAS	39 deg	39 deg
Ramp breakover angle	150 deg	153 deg
Extended profile, ECAS	145 deg	147 deg
Departure angle	30 deg	30 deg
Extended profile, ECAS	33 deg	33 deg
Standard wading depth	19.7in (500mm)	19.7in (500mm)

11 Competition and Adventure

Right from the very beginning of its career, the Range Rover has been looked upon as an ideal vehicle for certain types of competitive event. Over the years, it has taken part in international long-distance rallies and in off-road racing at club level, and has distinguished itself in each of these fields. However, where other manufacturers have committed hundreds of thousands of pounds to supporting 'works' competition vehicles, neither British Leyland nor Land Rover Ltd has ever been committed to a full factory-sponsored competitions programme for the vehicle. As a result, its successes have been recorded by privateers, some of whom have enjoyed a limited amount of factory support.

However, the manufacturers were very much involved in the very first plan to enter Range Rovers for competition, which was conceived back in 1969 before the vehicle had even entered production! Hugely impressed with the prototypes then running and with the paper specification, British Leyland decided to enter a team of three Range Rovers in the London-to-Mexico rally which was planned for the early months of 1970. One vehicle (apparently prototype number 3, AGN 316 G) was therefore shipped down to BL Motorsport at Abingdon for evaluation, probably some time in the autumn of 1969.

Unfortunately, the project went no further. Time was already tight, and it must have become clear very quickly that there would simply not be enough Range Rovers around for three production vehicles to be prepared for the London-to-Mexico event. As it was, the very first pilot-build Range Rovers were not put together until December 1969, and by this stage British Leyland had decided to enter a team of three Austin Maxis instead.

LONG-DISTANCE RALLIES

During the early 1970s, BL Motorsport was far too busy with other projects to return to the Range Rover as a potential competitions vehicle. Besides, the Range Rover was selling well enough not to need the favourable publicity which comes from a successful motorsport campaign; far more important for British Leyland at this stage was to secure the right sort of publicity for some of its lame-duck marques.

So it was left to privateers to realize the vehicle's potential and, as so often in off-road competition, it was the French who led the way. A number of Range Rovers were entered in the Abidjan to Nice Rally held in January 1976, and they claimed not only first and second places but also a further three places among the first ten vehicles to finish. Typical of a French-style 'raid', this rally was an enormously demanding event which lasted for seventeen days and covered 6,250 miles, and the Range Rovers' success in it was the first of many to come.

The next major event came two years later, and this time it was the Australians who were behind it. Leyland Australia decided to enter a Range Rover in the 1978 London-to-Sydney Marathon, a 17,000-mile (27,000km) endurance event of which a substantial element was to be run in the harsh conditions of the Australian outback. The rally vehicle and a support Range Rover were prepared by Del Force Automotive Services in Australia, by a pair of mechanics who had worked on Leyland Australia's 1974 World Cup Rally saloon car entrant. The same two drivers were engaged – Evan Green (father of *Car* magazine editor Gavin Green) and John Bryson, who secured sponsorship from his employers, Endrust.

The rallying Range Rover had a 4.4-litre V8 engine (an adaptation of the standard 3.5-litre type developed for Leyland Australia's P76 saloon), a lightened bodyshell, uprated brakes, auxiliary fuel tanks and many other extras. The vehicle peformed well on the desert sections of the rally, as Green had expected it to, but it was too heavy to be competitive in some of the others. Nevertheless, it achieved a very creditable class win and finished in tenth place overall.

It was events like these which persuaded the rallying fraternity at large that the Range Rover might make a good competition vehicle, but there were other factors also working in its favour. Not least of these was that the long-distance events of the World Rally Championship were becoming tougher towards the end of the 1970s, and so the entrants started thinking about using tougher vehicles. Among them, the Range Rover seemed to offer a good deal of potential, even though its weight looked like being a serious handicap.

So it was that 1979 saw the first-ever appearance of Range Rovers in a World Championship Rally, when the French drivers H. Cochin and C. de Rouvre used privately-prepared examples in the Bandama Rally. They were not successful, but from now on, the Range Rover would be an increasingly familiar sight in long-distance rallying. By far the majority of entries would be driven by French crews and prepared in France.

However, in the early 1980s it was in the Safari Rally, held in East Africa, that the Range Rover demonstrated its mettle. Privateer Rob Collinge took fifteenth place overall in the 1980 event, finishing first among the four-wheel drive entrants. Collinge and co-driver Mike Frazer again did extremely well with a Range Rover in the 1982 Safari Rally, this time claiming a class win and sixth place overall in a locally assembled vehicle supported by Cooper Motor Corporation. However, Collinge was not so lucky in the 1983 event, when his Range Rover crashed out of the event three sections before the finish. For 1984, the rules governing the Safari changed, and Collinge's vehicle would no longer have been competitive, despite an engine which had already been developed to give a reliable 200bhp with twin SU carburettors.

Meanwhile, the Range Rover had also begun to figure largely in the Paris–Dakar Rally, a punishing 6,000-mile (9,500km) high-speed event organized by a French group. The Paris–Dakar included classes for all types of vehicle – there were motorcycles, cars, specials, 4 × 4s and heavy trucks, to name just a few – and included special stages as well as stretches taken at maximum speed across open desert. The Ladies' Prize in the 1980 event went to Christine Beckers, who drove a lightened Range Rover; and there were several other modified Range Rovers among the entrants, as well as a British vehicle prepared by BL France.

Long-distance rallying often involves the service crew in making emergency repairs to a vehicle miles from anywhere. The large tent at least afforded some cover in this case.

For 1981, Land Rover also took an interest in the event, although their involvement was fairly low-key. They provided a Range Rover (registered TRW425R) for Tony Howard and John Miles of *Autocar* magazine to drive in the event, and tuning specialists Janspeed were also involved in preparing the vehicle. However, the honours went to BL France, who once again sponsored a Range Rover entry, this time in conjunction with VSD magazine.

The BL France entry for the 1981 Paris–Dakar in fact consisted of two Range Rovers, both prepared at Tom Walkinshaw Racing in Britain by mechanics from BL Motorsport at Abingdon and from conversion specialists Schuler Presses. Among the drivers engaged was French saloon car racer René Metge, who had competed in the previous year's event in a Leyland Marathon truck. This year, he and his co-driver Robert Giroux fully justified BL France's faith in them and their vehicle by reaching Dakar in first place overall.

By now, the Paris–Dakar event was beginning to attract serious participation from heavily funded factory teams, because it was attracting a great deal of media attention. Land Rover were forced to decide whether they, too, should pour money into a works competition team, but in the event they decided against it. The main reason, probably, was that sponsorship of the Camel Trophy event looked as if it would be more beneficial in terms of publicity. Nevertheless, BL France were authorized to continue their involvement with the event, and over the next few years it was once again the French arm of the company which supported the most successful rallying Range Rovers.

What was by now known as Austin Rover France (the name had changed in 1982) put its weight behind the French Halt-Up! rally team, and over the next few years the Paris–Dakar Rally saw a number of promising Range Rover entries from this combination. The 1984 entry was particularly successful, finishing in second place overall. That and other encouraging results were enough for Land Rover to look once again at becoming directly involved during 1986, and the Halt-Up! Range Rovers for the 1987 event were prepared with considerable technical assistance from Solihull. This resulted in an even better showing: a

Halt-Up! Range Rover driven by Patrick Zaniroli claimed second place overall in that year's event.

Typical of the specially-prepared Range Rovers used in long-distance rallying is this one, prepared in France by ETT in 1987.

Privateers meanwhile continued to approach the Paris–Dakar with determination. Among the unsuccessful entrants in the 1987 event was a heavily-modified Range Rover prepared by the Toleman Group (of Grand Prix and Formula 2 racing fame) and featuring body panels and floor pan made from the lightweight Kevlar now becoming common in Grand Prix cars. The vehicle was driven by Ted Toleman and Barry Lee, but despite a promising 290bhp engine prepared by JE Motors and some support from Land Rover Parts and Equipment, it broke both its axles and was forced to retire. Toleman did not compete in a Range Rover again, but he maintained his link with JE Motors, who provided the engine for the Toleman Group's 1989 Enduro T89 special as well as a 4.2-litre V8 for the Range Rover back-up vehicle.

As the Toleman initiative showed, the emphasis was now moving towards incorporating racing technology in the Paris–Dakar Range Rovers. For the 1988 event, the Halt-Up! team developed a new aerodynamic body – only barely recognizable as sharing some styling features with a standard Range Rover – and built it from lightweight Kevlar and carbon fibre. Halt-Up! also drew on the expertise of JE Motors for their engines, using 4.2-litre developments of the V8 which sat further back in the chassis than standard in order to improve weight distribution. These vehicles, allegedly capable of 125mph (200km/h) in rough desert conditions, earned third place for Patrick Tambay and fourth place for Malcolm Smith on the 1988 event.

Once again, Land Rover decided it might be time to consider direct factory involvement in a rally competitions programme, and during 1988 initial plans were made. In January 1989, work began on a feasibility prototype, intended to discover whether a rally vehicle designed to meet FISA Group T regulations would be competitive. The prototype (registered F 201 NWK) started life as a right-hand drive four-door with manual transmission, but was extensively modified with a JE Motors 3.9-litre engine capable of 215bhp and 247lb/ft of torque at 3,500rpm.

The vehicle's main bodywork remained unaltered, but it was fitted among other things with a Safety Devices internal roll cage, a huge fuel tank and a Torsen limited-slip differential in the rear axle. The resulting vehicle was allegedly capable of 140mph (225km/h), although the specification suggests that figure might be fanciful, but once again Land Rover decided not to go ahead. As usual, there were good reasons: the Discovery was about to be launched in November 1989; the Range Rover was about to take a further step upmarket and into a market sector where

ETT also prepared this Range Rover as a desert racer for Camel Racing Services. Its wheelbase was stretched by 10 inches (254mm) to improve high-speed stability.

competition prowess was irrelevant; and the factory had enough competitions activity on its hands with the Camel Trophy. Since then, there have been no further plans for the factory to take the Range Rover into motorsport events of any kind.

Nevertheless, the Range Rover's career as a desert racer continued into the 1990s, thanks primarily to French interest. Among the French companies involved in the preparation of Range Rovers for long-distance rallying, the best known is undoubtedly ETT, who have been been preparing Land Rovers and Range Rovers for events like these since the late 1970s. The company's Range Rovers have competed in a wide variety of of desert rallies, including the Paris–Dakar, the Atlas Rally, the Aragon Baja and the Paris–Peking, and took top honours in all of them during the 1980s.

The initials ETT stand for Eggenspiller Tout Terrain, the last two words being French for off-road and the first being the surname of the company's founder, Christian Eggenspiller. He is a passionate believer in the Range Rover's abilities as a rally car, and he has used his twenty-odd years of experience to produce what are perhaps the most well-prepared off-road Range Rovers to date. Among them have been several vehicles for the Camel Racing team, and two of these finished second and tenth overall in the 1987 Paris–Dakar rally.

Every Range Rover to receive the full ETT treatment is stripped right down to its chassis, and almost all of them are then stretched by 10in (254mm). Eggenspiller

Late two-door models are favoured for off-road competition in France, but such vehicles were not available in the Range Rover's home country.

argues that the longer wheelbase gives greater straight-line stability at high speed on sand, and therefore few of his customers have not taken up this option.

After stretching the chassis, ETT then replace all of the Range Rover's body panels with extremely light (but expensive) Carbon Kevlar equivalents that were designed by Eggenspiller himself. As well as giving significant weight saving, carbon Kevlar is immensely strong and is therefore ideally suited to the tough rally environment. The body panels themselves retain the Range Rover's basic shape, although the bonnet, front wings and grille panel can be made as a single piece to produce a lift-off panel which gives complete access to the engine bay.

To give superior performance for the 100mph-plus requirements of desert racing, the V8 engine is also heavily modified to produce in excess of 250bhp. This is achieved by extending the cubic capacity to 4.8 litres and by modifying the ECU to inject more fuel into the engine. Roll cages, bucket seats, underside protection and modified suspension are the last additions to an ETT rally Range Rover.

Not every ETT Range Rover is modified to this extent, however. For clients with more modest budgets, ETT produce the Marathon series. This off-road package includes a full roll cage, an aluminium front skid plate, welded-on differential protectors, long-range fuel tanks (as large as 15 gallons/70 litres), internal lighting, Terratrips, internal body and chassis strengthening, uprated springs, twin dampers and bucket seats with four-way harnesses. Priced at up to half the cost of a brand-new basic-specification Range Rover, such conversions are not cheap – but then the Marathon modifications do boast an impeccable pedigree, having been proven in the world's most arduous desert rallies.

CAMEL TROPHY

The Camel Trophy is often seriously misunderstood. Although it contains an element of competition, it is not a competitive motorsport event in the same way as the Paris–Dakar Rally. It is more of a combination of expedition and adventure, in which teams from different countries negotiate tracks and trails through inhospitable terrain in a loose convoy. Its aim is for all the vehicles and all the teams to get through, and to that end the 'competitors' actually help one another instead of trying to do one another down. There is a first prize, for the Camel Trophy itself is awarded to the team which has scored most points in the expedition section and in a series of Special Tasks set during the course of the event; but equally important is the Team Spirit Award, which goes to the team which makes the greatest contribution to the overall success and achievement of the event.

The first Camel Trophy event was run in 1980 with the backing of R. J. Reynolds Tobacco GmbH, the German makers of Camel cigarettes. Just three teams took part, all from West Germany and all driving Jeeps, and the event was a 1,000-mile (1,600km) driving adventure through the jungles of the Amazon basin. However, the nature of the event changed for 1981: Land Rover agreed to supply the vehicles and the technical support, and from 1982 the Trophy was opened to international teams. Since then, the Camel Trophy has gone on to become a respected event in its own right, often performing tasks of environmental importance and helping local communities as it passes on its way.

Range Rovers were used on three Camel Trophy events in the 1980s, in each case being standard production-line models kitted out with additional equipment by the Special Projects division (known as Special

Adjustments in mid-rally are of course necessary and once again the circumstances are often far from ideal.

Vehicle Operations after July 1985 and in the 1990s as Land Rover Special Vehicles). This equipment included a heavy-duty roof rack on which were mounted additional driving lamps, a heavy-duty winch, skid plates under the vehicle and a full roll cage inside, and special off-road tyres. In addition, the Range Rovers of course carried a large quantity of expedition equipment: jerrycans, ropes, tools, high-lift jacks, additional spare wheels, and so on.

On the 1981 Camel Trophy in Sumatra and the 1982 event in Papua, New Guinea, Land Rover provided Range Rovers for all the national teams. The vehicles were of course 3.5-litre carburettor models with manual transmissions, and in both cases two-door models were used. However, on the third and last Camel Trophy event to feature Range Rovers, the 1987 event in Madagascar, the fourteen national teams drove four-door models with the 2.4-litre VM turbodiesel engine. Naturally, the demanding nature of the terrain on all of these events caused failures of axles, differentials, clutches and even suspension components, and the Range Rovers suffered more than their fair share of body damage. Yet, in the end all of the vehicles got through to their destinations – a tribute to the rugged strength and rough-terrain abilities of the Solihull flagship.

OFF-ROAD RACING

For the first half of the 1970s, Range Rovers were too new and too much in demand for the off-road trials fraternity to get to grips with them. There was no doubt

The 1987 Camel Trophy, held in Madagascar, featured the Range Rover Turbo D.

that they were able off the road, but few people were prepared to risk damaging an expensive new Range Rover in a trial when they could have just as much fun and be almost as competitive in an elderly leaf-sprung Land Rover which had cost a fraction of the price. This led to a rather ridiculous attitude in certain sections of the off-road driving fraternity, where Land Rover die-hards argued that a vehicle which had been selected for display at the Louvre as a modern work of art could hardly be taken seriously for demanding outdoor work. Such a vehicle could not possibly be as rugged as a Land Rover.

However, the Range Rover had won its off-road spurs early on, even though it did not become a popular choice in motorport events until much later. The event was the Welsh Hill Rally of May 1971, when no fewer than four examples were entered. Among them was pre-production vehicle number 23 (registered YVB166H), borrowed from Solihull for the occasion by Roger Crathorne, who had been one of the development engineers on the Range Rover project. Crathorne was no mean driver off-road (much later, he took charge of Land Rover's off-road driving school and demonstration team), and he succeeded in bringing the vehicle home in first place. To demonstrate that this was not simply a case of superior driving skill, however, two of the other three Range Rovers entered for the event took fourth and fifth places behind him.

The position began to change in the mid-1970s, when the earliest Range Rovers had passed on to second, third and fourth owners, and Range Rover ownership was no longer the exclusive preserve of the wealthy. A few Range Rovers began to turn up at All-Wheel Drive Club events in Britain and at regional events organized by the Land Rover clubs. And at about this time, the first Range Rover-based 'specials' began to appear.

Off-road trials often involve negotiating course markers as well as the difficulties of the terrain itself.

The most successful of the specials builders was garage proprietor Alvin Smith, who started by winning a number of club competitions in the UK with his specially-prepared Range Rover. In addition, he took the vehicle to France and came fourth overall in the gruelling Rallye Infernal. HWA 576 J was an early model, modified for greater power with high-compression pistons and an Iskenderian camshaft originally designed for the V8 engine's Buick ancestor. Suspension reinforcements helped it to stay together at high speeds in the rough, and to reduce body roll around corners, the top half of the rear body had been removed to create a pick-up configuration. Alvin Smith later went on to use the knowledge he gained with this vehicle to build a series of ever more astonishing Range Rover-based off-road racers.

The earliest was the Strange Rover, which was developed with some assistance from Range Rover conversion specialists Schuler Presses, and used modified Range Rover components in a Triumph TR7 bodyshell. So successful was this that Alvin Smith went on in the early 1980s to make a 'production' version, the Strange Rover 2, which was suitable for minimal bodywork in the best off-road trials tradition and had its engine mounted amidships.

As the 1970s turned into the 1980s and the popularity of off-road motorsport increased, so increasing numbers of older Range Rovers which had fallen on hard times were converted into off-road racers. A favourite method of lightening these vehicles was to cut off the upper rear body panels to make a pick-up configuration; engine modifications also increased power and torque to make the Range Rovers more competitive. For those classes where such modifications were permitted, vehicles were also 'bob-tailed' (that is, the rear overhang was shortened to improve the departure angle), and some Range Rover chassis were shortened and fitted with Land Rover bodywork to produce extremely able off-road hybrids.

FRENCH ENDURANCE RALLYING

In France, however, the off-road racing scene was once again different. While off-road racing in the UK has always been the

Trials specials come in all shapes and sizes. This is a 'bob-tailed' Range Rover, with the rear overhang shortened to improve the departure angle...

... this is one of Alvin Smith's Strange Rovers, based on a Range Rover chassis and running gear...

... and this is a typical hybrid, with Land Rover body panels on a shortened Range Rover chassis.

preserve of private enthusiasts with – it must be said – quite limited budgets, the French take a very different approach. Over there, off-road racing means endurance rallies, and such events take place almost every month, attracting full works support from all the major manufacturers. Among them, of course, is the French importer of Land Rover products, now known as Rover France.

Rover France became involved with the French Endurance Series in 1993, when they were approached by two young French drivers, Frederick Schotts and Stéphane Barbry, with a view to setting up an official Rover team. Prior to 1993, Schotts and Barbry had been racing their own privately-prepared Range Rover very successfully, but they now felt that they needed official support if they were to carry on.

After looking at their proposal, Rover France supplied them with a pair of two-door Range Rovers to the current 3.9-litre specification, and these were then turned over for preparation to Arras Racing. As the Range Rovers were to compete in the T1 category (for standard vehicles), preparation was limited to safety equipment only, and the engine and transmission remained standard. As some endurance rallies can last for up to twenty-four hours, this speaks volumes for the faith that Land Rover put in their products' ability and durability.

This faith was well-founded, because Schotts and Barbry came a creditable second overall in the endurance rally series in their first season with Rover France. And 1994 was to prove even more successful, with the official yellow and green Range Rovers taking first place overall, beating the rival factory teams of Nissan, Mitsubishi and Toyota, all of which had larger budgets and modified, T2-class, vehicles. It was unfortunate that the focus of publicity was on the second-generation Range Rover by the end of 1994, otherwise Land Rover would surely have made more of this astounding result. As it was, few people outside France can have been aware of it.

Apart from the official Rover France works team, the Range Rover remains popular with private French teams for endurance rallying. Each team spends large amounts of money on its off-road racing, with most privateers entering vehicles that are no more than two years old. These vehicles are used exclusively for racing, and normally have to be rebuilt after every arduous 24-hour event.

The Rover France crews won the French Endurance Rally Series in 1994 with their specially-prepared Range Rovers.

The majority of private competitors in endurance events race highly-modified two-door Range Rovers, which were of course freely available in Europe until early 1994. These have usually been completely rebuilt to make them more capable and robust. Twin dampers are common, as are tuned 3.9-litre engines, strong gearboxes from the Spanish Santana company (who supplied them as standard equipment for early Land Rover 110s), full roll cages and racing seats.

Sadly, the Range Rover did not figure in Rover France's endurance racing team for 1995. Nor indeed did its successor, the second-generation Range Rover introduced in 1994, as Solihull would not sanction its appearance in these competitive events. Instead, Rover France competed with a Land Rover Discovery which had a John Eales 4.0-litre V8 engine under its bonnet. Nevertheless, the 'old' model had gone out in a blaze of glory after winning the 1994 series, and will no doubt continue to be used by privateers for some time to come.

THE DARIEN GAP EXPEDITION

Solihull never had any trouble finding customers for the Range Rover, although in the vehicle's early days there was a persistent suspicion among the uninformed public that this semi-luxury vehicle could not possibly have the rugged off-road abilities of a Land Rover. While this did not seem to be hindering sales, it was certainly an impression which British Leyland was anxious to counter, and it may well have been for this reason that the company agreed to supply the British Army with a pair of Range Rovers to use on the 1971–2 British Trans Americas Expedition. This expedition was later known more commonly as the Darien Gap Expedition after its most famous and most difficult element.

In fact, the Darien Gap was central to the expedition. It was a 250-mile (400km) stretch of swamp and impenetrable jungle which prevented road traffic passing between the north and south parts of the American continent. As such, it was a barrier to trade and development in the area, and local action committees had been set up whose objectives included persuading the local authorities to build a road through the Gap. To gather publicity for their aims, these committees proposed that an expedition should attempt to cross the Darien Gap using road vehicles. In Britain, their ideas were espoused by the British Trans Americas Committee, and from there the idea that Britons should mount this expedition was passed to British Leyland and to the British Army. Both accepted the challenge with enthusiasm.

Before the expedition got under way, the original idea of a 250-mile Darien Gap crossing had been superseded by a plan for a British Army team to drive two Range

The crew and vehicles of the 1971–1972 British Trans Americas Expedition before setting off. The vehicle on the left is wearing the huge swamp tyres which were to cause differential failures in the swampy terrain of the Darien Gap.

Rovers the full 18,000-mile (29,000km) length of the American continent, from Alaska in the north to Cape Horn in the south. It was an ideal arrangement, which allowed the Range Rovers to demonstrate both durability and ruggedness, while drawing attention to the fact that the Darien Gap was the last remaining obstacle to completion of the Pan American Highway connecting North and South America. Lastly, a scientific programme was also added into the expedition, aiming to study the botany, biology, geology, geography, entomology and zoology of the Darien Gap region.

The two Range Rovers chosen were early left-hand drive examples, registered VXC765K and VXC868K, and they were prepared at Solihull towards the end of 1971. Although their engines and transmissions remained standard, they were equipped with special features like prototype front seat head restraints and rear window wash/wipe systems. In place of the rear bench seat, each vehicle had a single reclining Rover P6 saloon seat, to allow off-duty crew members to relax in comfort.

The two vehicles set out from Anchorage, Alaska, on 3 December 1971, and completed the run down the roads of the North American sub-continent without difficulty. However, problems occurred when they reached the Darien Gap. Although the expedition had been planned to cross the area in the dry season, the tropical rains were late and the jungle and swamp were at their most treacherous when the expedition reached Panama City in January 1972. The wide swamp tyres fitted for the Darien Gap crossing put enormous strains on axles and half-shafts when used in the boggy ground, with the result that both vehicles suffered transmission failures.

New parts were flown out from Solihull, and engineer Geof Miller was sent with them to keep an eye on things, but the only real answer was to reduce the strain on the vehicles by decreasing their payloads. So, instead of carrying their own supplies, the expedition had to rely on airdrops and pack ponies for the rest of their crossing. In addition, they decided to use a Land Rover as a pathfinder vehicle, and so a second-hand 88-inch was located in Panama and drafted into service.

The crossing proved more than ordinarily challenging, and the fact that it took ninety-nine days for the expedition to fight its way through 250 miles of swamp and jungle tells its own story. On many occasions, the Range Rovers were able to make progress only by inching their way along the lightweight aluminium ladders they carried, by crossing rivers aboard a special Avon inflatable raft, or by driving through gaps in vegetation which had been created by blasting. However, make it they did, emerging fom the jungle on 23 April 1972 for the straightforward run down to Ushuaia in Argentina, which they finally reached on 10 June.

The two Range Rovers used in the expedition still survive. One belongs to the Heritage Collection at Gaydon, and the other is on permanent loan to the Dunsfold Land Rover Trust. Expedition leader Major John Blashford-Snell wrote a fascinating account of the Darien crossing for a Range Rover publicity leaflet issued in September 1972 (number 844/9.72, *The Darien Crossing*), and the Army made a 32-minute promotional film (*Darien Conquest*) about it. No-one who has read the former or seen the latter can be in any doubt that this was the event which finally overcame prejudice that the Range Rover was not as capable or as rugged as a Land Rover.

12 A Range Rover Buyer's Guide

Unlike many vehicles which are the subject of enthusiast attention, the Range Rover is not and never has been a rare beast. More than a quarter of a million have been made in twenty-five years of production, with the result that there is a plentiful supply of second-hand examples in most of the territories where the vehicle has been sold. However, prices, specification and condition vary widely, and finding a Range Rover to suit your taste and budget may be somewhat more tricky than these statistics would suggest.

As with most modern classics, it is condition rather than age that dictates the price of second-hand (or, as some franchised dealers would have it, pre-owned) Range Rovers. It should be fairly obvious that the older the vehicle is, the greater are the odds that it will be in poor condition. It will therefore be much easier to find a good late vehicle than a good early one.

Early 1970s models in good-to-excellent condition are rare, for example, although it is possible to find good examples with a little searching. Early models that have undergone restoration are more easy to find than originals, so low-mileage versions of the latter can command quite high prices, while tired examples crying out for a chassis-up rebuild will sell for a few hundred pounds.

This chapter aims to help you make an informed choice when buying a second-hand Range Rover. As well as giving some handy hints on what goes wrong and how to spot it, we shall also offer advice on the best models to go for, depending on your intended use and budget.

BUDGETS

Setting a budget is crucial before you buy a Range Rover. Major faults can be very expensive to rectify on both the simple early vehicles and the later examples with their more complicated electronic systems and you need to be quite certain how much you can afford. The authors' experience suggests that you should work out your total budget for the purchase plus first year of ownership and then deduct 10 per cent. This will then give you a sensible purchase price and allow for repairs to faults which were not apparent at the time of purchase.

The early chapters in this book will tell you what is available in terms of specifications. For example, if you want a right-hand drive two-door, you will be unlikely to find anything available in the UK market registered after 1986, apart from the CSK special edition. Similarly, although early four-door Range Rovers are comparatively cheap to purchase, we would recommend spending a little extra on a later vehicle in order to have the benefits of the five-speed manual or four-speed automatic gearboxes.

Generally speaking, you get what you

pay for with a second-hand Range Rover. A Range Rover bought for less than £1,000 in the UK is unlikely to be fit for anything other than trialling, as the costs you will inevitably incur to get the vehicle looking good or (more importantly) through the MoT test are likely to be greater than the true value of the refurbished vehicle. Nevertheless, if you are a competent DIY mechanic with plenty of spare time, a Range Rover restoration project could be an entertaining and interesting idea. It all depends on your own budget and mechanical knowledge.

As has been proven by many enthusiasts (including both the authors with their own Range Rovers), an early Range Rover can be bought and then modified using late-specification parts and accessories. The Range Rover, like other vehicles from the Solihull stable, is basically a huge Meccano set, so upgrading the interior, transmission, engine and bodywork is simply a matter of locating the parts for the right price. James' Range Rover, for example, is based on a 1973 chassis and sports post-1987 four-door bodywork, a 200Tdi engine, five-speed gearbox with chain-driven transfer box and a late-specification Vogue interior. Nick's 1978 Range Rover has had new sills, inner wings, upper and lower tailgates, an aftermarket Nationwide Trim interior and a complete re-spray. The V8 engine, likewise, has also been completely overhauled, as has much of the transmission.

With both the authors' Range Rovers, second-hand parts (especially bodywork) have been used extensively. There is a large market in *bona fide* Range Rover breaking and magazines such as *Land Rover Owner* list the appropriate suppliers, with some companies specializing in second-hand parts almost exclusively from crash-damaged late Range Rovers.

WHICH MODEL?

The huge variety of factory specifications and the vast array of aftermarket accessories mean that it is rare to find two Range Rovers which are exactly alike. You should always start your search for a Range Rover by asking yourself some basic questions. Do you want two or four doors? Manual or automatic transmission? Petrol or diesel? Carburettor petrol engine or injected type? Vogue or Vogue SE? The choices are many, even before you reach details like the ideal body colour!

It will help you to answer these questions and others if you ask yourself what exactly you plan to do with your Range Rover. If it is to be used as your main car, you will need to look at reliable, tidy machines; possibly with four doors if you have a family. On the other hand, if the Range Rover is primarily to be used as a weekend fun car, possibly for off-roading, then it might be worth considering a vehicle that is solid on the inside, but tatty externally. If the vehicle is not being used all of the time, you will be able to restore it as time and money permits.

Once again, budget will play a large part in your ultimate choice, although our advice would always be to go for the best vehicle that you can afford. With a more expensive vehicle, less should need doing to it in the short term, thus allowing you to overcome the initial shock of the purchase and to prepare yourself for the first item that will need attention!

As the Range Rover is a vehicle that is as much at home on the motorway as it is up to its axles in mud, you should be especially careful of vehicles that have had a hard life. Late-model vehicles that have spent their entire lives in the fast lane of a motorway are likely to be as tired as older ones that have been pottering around on a farm.

In both cases, the magic letters FSH (for Full Service History) should put your mind at rest, for if the vehicle has been excellently maintained, a Range Rover will take high mileages and driving around in mud in its stride.

The adaptable all-purpose image of the Range Rover also means that many vehicles have been modified with accessories or, indeed, customized to the (often) extravagant whims of the owner. In many respects, a vehicle modified in this way offers excellent value for money, as accessories and conversions rarely affect second-hand prices and you are therefore getting something for nothing. However, make sure that you do want the accessories attached to a vehicle: if you do not want them and therefore do not use them, they will probably have ceased to work by the time you come to sell the vehicle on, and that will certainly reduce buyer interest and the price you can ask.

Do be especially cautious about modifications that have been carried out by incompetent DIYers, as wiring looms overheat and fuses blow at the most inconvenient times when they are overloaded by extra lights or electronic gadgets. Similarly, all add-ons should have been safely and sympathetically added. Does the overdrive operate without crunching? Does the winch work properly? Are the side steps bent? And most importantly, how much will they cost to fix?

Aftermarket diesel conversions are an interesting modification to consider, as a good Japanese engine will add price to a tatty vehicle, although a slow Perkins will devalue an otherwise smart one. In every case, check that the engine has been professionally fitted – if not, think twice before buying.

As a rule of thumb, ignore naturally-aspirated diesel conversions and go for turbocharged types; intercooled ones are even better. Safest bet of all is the factory's own Tdi conversion (available as a kit since 1993), but among the better engines from other manufacturers are turbocharged types from Mazda, Nissan and Isuzu. All of these offer performance similar to that of the carburetted V8. On the other hand, Range Rovers fitted with low-revving, naturally-aspirated Ford York or Perkins diesels will be lucky to reach a very noisy 60mph (100km/h) downhill; only gearing revisions (which are comparatively rare) can help improve maximum speed and they inevitably have a negative effect on acceleration.

Specialist conversions are not common but can offer spectacular value for money – especially if the owner has no idea of their rarity value. However, some conversions (particularly those with a lot of bodywork modifications) can become very tatty after only a few years' use, with replacement parts and the knowledge required for repair limited. It is also important to remember that custom-built vehicles were usually built to meet the needs and tastes of an individual and that these needs and tastes may not be identical with your own.

With six-wheelers and extended-chassis conversions, pay special attention to the chassis and areas that have been cut and re-welded. Although most of the specialist conversion companies had very high standards, there were some less scrupulous firms which produced very inconsistent work. Replacing one-off panels could also prove to be very expensive, while finding bodywork specialists with the necessary ability to adapt what already exists may also prove tricky.

Even the factory-approved Monteverdi four-door conversion is far from straightforward to repair or restore. The converted bodies are are especially prone to rust on the rear wheel arch just below the rear door and there is no easy way to repair this because all stocks of spare panels were exhausted many years ago. Particularly bad rust is likely to result in the vehicle failing its MoT, so beware of patched repairs and 'bodged' jobs. The Monteverdi's luxurious interior will also be very expensive to restore if it is damaged, because the only way to replace some items will be to have them re-made by hand.

Beware of problems on vehicles with special bodywork. This is the rust-prone rear wheelarch of a Monteverdi four-door conversion – a panel which is no longer available.

However, all varieties of Range Rover, whether production standard or heavily modified, have certain common problem areas which you should investigate before buying. What follows is a guide to what you should look for.

BODY AND BODYFRAME

Like the Land Rover that went before it, the Range Rover has outer body panels made of corrosion-resistant aluminium alloy. However, not all the Range Rover's skin panels are alloy: the rear tailgate and bonnet are made from steel, and rust does manifest itself on these items all too frequently. In addition, the inner body frame is a steel skeleton and that too can rust in quite spectacular fashion.

It is very unusual for a Range Rover that is more than ten years old not to have any rust on the lower tailgate. Typically, this panel rusts from the bottom upwards and some rough examples actually come away from their internal mountings. Once the rust has taken hold, filler and touch-up paint are a waste of time; the only option is complete replacement. The upper tailgate also rusts, particularly in the bottom corners along a welded seam. From the mid-1980s, these seams were protected by plastic covers and are less prone to corrosion. Early ones can deteriorate very badly, though, and once again the only remedy is complete replacement of the upper tailgate, complete with its glass.

Bonnets on pre-1980 models tend to rust around the holes for the badging, and stone chips in this area on bonnets of all ages can also lead to rust. The leading corners of the bonnet tend to rust from the inside out and are very difficult to repair once they have holed. Bonnets are also often damaged in

The lower tailgate is steel and rusts badly.

Most of the visible panels on a Range Rover may be of aluminium alloy, but they still corrode. The white patches here show typical corrosion damage where it regularly occurs – on the flange where the alloy outer front wing bolts to the steel inner.

accidents and the cost of a replacement panel is high enough to encourage most owners to go for a quick patch repair with filler. Unfortunately, such repairs rarely last long; the shocks caused to them when the bonnet is slammed shut tend to make them crack out after a while. For all these reasons, it is advisable to avoid a vehicle with a damaged or repaired bonnet, or at least to budget for the cost of a replacement panel.

The inner bodyframe is visible in certain places and those places are likely to have been attacked by rust. Two notorious rust-spots are around the base of the A-pillars (the front door hinge pillars) and on the rear cross-member. Although the door pillars can be patched up with steel plates, the rear cross-member has to be completely replaced for the vehicle to pass the MoT because it is essential to the structural integrity of the vehicle. Beware of poor welding, and look also for unscrupulous filling that has no structural value.

Although aluminium alloy is corrosion-resistant, it is very far from being corrosion-proof. Range Rover body panels suffer from a type of aluminium corrosion caused by electrolytic reaction when the metal comes into contact with the steel frame. This shows up as white powdery areas, initially underneath bubbling or flaky paint. In vehicles that have been particularly badly affected, the corrosion will actually hole the body. This is most common along the top mounting flange of the outer front wings, where the alloy panel bolts on to the steel inner wing underneath the bonnet. Other places to look for aluminium corrosion are the bottoms of the doors, where the alloy panels wrap over steel frames.

You should also look underneath for body problems. Using a torch, shine a light up into the sills, as these are steel and can also rust. Even though the Range Rover has a separate chassis, these sills are structural: weak ones will cause the body to sag and will make the vehicle unsafe. Replacing sills is a long and tedious job, complicated by the need to strip out most of the interior, so avoid Range Rovers with poor sills. Remember to examine the sill

A typical example of rust in the front of the steel sills which form part of the inner bodyframe.

carefully if it is caked in mud, as this can hide all but the very worst corrosion. If you really cannot be sure about the condition of a sill, tap it with a small hammer. If it sounds solid, all is well; if you go through the metal, walk away from the car!

The rest of the underside of the vehicle should be fairly solid, with the only other problem areas being the front footwells on the bulkhead and the front and rear inner wings. The footwells can be inspected from inside, although you will have to lift up the floor covering to see the metal. That floor covering may have as many as four layers – overmat, carpet, rubber mat and aftermarket soundproofing – so the job may not be as straightforward as it sounds! Front footwells tend to corrode at the corners, where mud has accumulated from off-roading. Small holes can usually be plated, but big holes or large areas of corrosion lacking good metal to weld onto are a problem. Aftermarket footwell panels are available, but it is neither easy nor cheap to make a good job of fitting them.

The steel inner front wings can be inspected after opening the bonnet. If the vehicle has seen regular off-road use, these panels are among the first to be affected by corrosion, so look carefully. On bad examples you may be able to see through to the ground, but providing you have enough metal to weld onto, plates or repair sections will make an adequate short-term repair. Bad examples will need the inner wings replacing, which is again a labour-intensive job.

The steel footwells rust and eventually perforate. On this example, the front wing has been removed to allow access, as the whole footwell is to be replaced.

Rear inner wings are less prone to corrosion than the front ones. However, they can be exceptionally bad, especially around the

The rear inner wings are also steel and will eventually corrode, particularly around the wheelarch area.

join between inner wing and wheelarch. Another weak spot that can act as a guide to the general condition of the rear inner wings is the recess under the spare wheel. If it has rusted badly (it can actually hole, too), the inner wings are likely to be in a similar state.

The last area of bodywork to check is the doors. Two-door models are particularly vulnerable to problems here. Their big doors, weighed down by frame and glass, are the heaviest single body components on a Range Rover, and their weight will pull against A-pillars and body sills weakened by corrosion. This eventually causes them to sag, to such an extent that the top of the door hits the 'B' pillar every time it shuts. The only cure for this is a replacement sill, complete with a re-welded 'A' pillar – a very expensive exercise.

Remaining with the doors, front quarter-lights tend to rust badly at the base and in bad cases this will let water in when it rains. Replacing the quarter-lights is unfortunately not an easy task, as you have to replace the whole top frame, which means taking the door to pieces. Most owners learn to live with the slight leak after a while.

The door mirrors used in the late 1970s and early 1980s are particularly prone to rust and can look dreadful, like this.

CHASSIS, AXLES, SUSPENSION AND WHEELS

The robust steel ladder frame chassis has always been one of the Range Rover's strongest points, both literally and metaphorically. Rusty chassis are fairly uncommon, although corrosion can manifest itself on the curved rear section behind the rear springs. Plating is usually possible, although it is a more complicated job than it appears at first sight because the fuel tank needs to be removed first for safety reasons. Such rust is normally caused by the steel coming into contact with salt water – a common problem if the vehicle is used to drag boats out of water. Accordingly, if you see a Range Rover with a boating club sticker affixed to the window, pay particular attention to the rear chassis.

Apart from this, the chassis should be fairly solid. However, while underneath the vehicle, check that the bracket supporting the rear bodyframe is in place. This is situated just above the rear springs, and can be seen as you look through the rear wheel-arch. If the bracket has rusted away, the body will be resting on the chassis itself and the structural integrity of the vehicle will be compromised.

Axles are robust components, but when looking under the vehicle you should check for signs of damage to the differential casings. This will almost invariably have been caused while off-roading, and you will be unable to tell easily whether there are further problems lurking inside the damaged casing. Check, too, for oil leaks from the ball swivels at the outer ends of the front axle (they have bright metal surfaces): serious leaks will lead to problems and to an unroadworthy vehicle. On later vehicles fitted with anti-roll bars, check that the brackets which attach the bars to the axles have not been bent: tyre fitters who do not look too closely where they put their trolley jacks can cause damage here.

The Range Rover's coil spring suspension is relatively straightforward to replace, but problems in a vehicle for sale can make good bargaining points. A common fault to look for is rust around the mounting flanges of the front damper turrets; heavy off-roading could cause a tower to tear away from its mounting. Fortunately, the turrets are inexpensive and fairly easy to replace. On the other hand, the Boge self-levelling unit is expensive to replace. While it generally wears well with age, oil leaks around the rubber boots are a bad sign and a leaky self-leveller should fail the MoT test.

There is no need to worry too much about sagging springs and tired dampers. However, a sagging rear is a good sign of a hard life towing, so think of the possible strain this may have placed on other components, particularly the transmission. On vehicles that sit low, wallow in corners and feel undersprung, it is amazing what a new set of dampers and springs will do. We would advise anyone selling a Range Rover to change the springs and dampers first because it will make a vehicle much nicer to drive and first impressions count! Similarly, buyers should know that soft suspension can be easily cured, but can always be a bargaining point.

The Range Rover's suspension bushes wear quickly if the vehicle is used a lot off-road and replacing them can be expensive. To spot worn bushes on the front, steer the front wheels from lock to lock while stationary. Ask someone outside to observe whether the side of the body and wheel centre stay aligned; if they do not, the bushes are worn. To check the rear ones, drive uphill on-road in third gear. With your hands gripping the steering wheel lightly in the straight-ahead position, put the power on. If the car pulls to the right, the rear triangular bushes are worn. When you back off on the power, the vehicle should regain its original straight line. If these tests do show up worn bushes, it may be worth considering aftermarket 'Polybushes' as replacements, as these do last longer than the standard type (although they also give a slightly firmer ride).

While on the test drive, check also for steering wobble at speed. Although this could just be the front tyres and wheels out of balance, it could also be badly worn wheel bearings or tired steering ball-joints. Wheel bearings expire with regular off-roading and these can be spotted by a 'groaning' noise while cornering. If you hear this, jack up the car and rotate each wheel individually. Each one should turn freely, although if the bearing has gone it will be stiff and make a grating sound.

Turning now to the wheels, the steel Rostyle versions that are a Range Rover trademark tend to rust on the surface, although second-hand versions can be picked up cheaply, or the originals can be restored with a sandblaster and some silver Hammerite. The Vogue alloys can also be affected by corrosion, which shows itself as the usual white powder, often underneath patches of flaky paint on wheels with painted centres. Watch out, too, for scuff marks from pavements, or even dents caused by off-roading or kerb-bashing. If you need to replace a Vogue alloy (or wish to change to them), be warned that the wheels are extremely expensive, even on the second-hand market.

ENGINE

The V8 petrol engines are all pretty solid, although the top end can cause problems on older vehicles that have not been regularly serviced. If regular oil changes have not been observed, camshafts tend to start wearing quickly at around 80,000 miles (130,000km). Camshaft wear can be spotted through noisy tappets or general top end chatter – especially when the engine is started from cold. On carburettor engines, uneven idling is likely to be caused by wear in the carburettors or in their linkages and this can be more costly to repair than it might initially appear. On injected engines, beware of all behavioural peculiarities: faults within the ECU (the engine's computer 'brain') are quite common and the unit cannot be repaired but must be replaced at considerable cost. Be wary in particular of vehicles with a 'G' model-year code (*see* table in Chapter 5), as many of these seem to have had faulty ECUs.

Most Range Rovers built before mid-1984 had an oil pressure gauge and you should know that this is an extremely unreliable instrument. Firstly, it is not a precision instrument anyway and secondly it can become faulty if mud damages the electrical contacts. The V8 was in any case designed to operate with relatively low oil pressure, so do not worry about what appear to be alarmingly low oil pressure readings unless there are other reasons to suspect the engine's health. Generally speaking, you only really have a problem if the oil light on the dashboard fails to go out, or if the engine sounds particularly rough.

Faults such as these are not too difficult to cure, as oil pumps can be changed, oilways flushed through and camshafts changed comfortably over a weekend. However, the cost of a new camshaft, a new set of hydraulic tappets and the gaskets necessary to complete the rebuild can easily exceed the cost of a healthy second-hand engine from a breaker's yard! Remember that the V8 engines from Rover saloons are similar in most respects to the Range Rover's V8 and will generally cost less from a non-specialist breaker. You may need to change various ancillaries, such as the water pump, oil pump adapter plate and carburettors, but many owners have carried out such engine swaps very successfully. Avoid the pre-1973 saloon engines with their 10.5:1 compression ratios, though: these were designed to run on high-octane petrol which is no longer available.

Like most diesel engines, the factory-fitted Tdi and VM engines are capable of many hundreds of thousands of miles, providing they are correctly maintained. However, some examples of the VM engines have persistent head gasket problems, so check the vehicle's service history. Regular oil changes are also essential on all these engines, as dirty oil or low oil levels will quickly lead to failure of the turbocharger bearings. Check for noise from the turbocharger (a whistle is normal) if the service history is unclear, and obviously check the condition of the oil and the oil level itself. Watch, too, for excessive amounts of black smoke emerging from the exhaust under acceleration. This can indicate cylinder head or piston ring problems.

Finally, after your test drive, park the vehicle on a clean section of road and leave it ticking over. After about five minutes, check underneath for oil or water leaks. Whereas the power steering system is notorious for spilling oil, the engine itself should be leak-free; if not, there are probably gasket problems which will need further examination.

TRANSMISSION AND BRAKES

Expect transmission whine on Range Rovers equipped with the four-speed LT95 gearbox; there never has been a cure. However, you will know that there is a fault with the gearbox if the whine is different or louder in one gear only. Gear changes always feel vague, but there should be no accompanying grinding noises. If there are, it is highly likely that there is excessive wear in the clutch or gearbox, or maybe both. Beware also of gears slipping out of drive on the overrun, something to which both four-speed and five-speed gearboxes are susceptible.

The Range Rover manual gearbox (especially the LT95), along with the rest of the drivetrain, is very solid. You can tell if it is worn by excessive clunking and juddering when you pull off in first gear, high ratio. If the gearbox is worn, walk away from the car, unless it is spectacularly cheap. New gearboxes are horrifyingly expensive and good second-hand ones are difficult to come by. Gearbox replacement is also not an easy job, necessitating removal of the interior and several hours' labour.

When testing the transmission, try engaging both high and low box on the transfer gearbox. They should engage easily, although lack of use may make the lever stiff. In a sense this is a good sign, as it means that the vehicle has not been regularly off-roaded.

Automatics tend to give years of good service, although be especially wary of leaks around the gearbox housing. The automatic shift should operate easily and the manual override should operate correctly in each gear. When selecting reverse, beware of metallic clonks or a sharp jolt as the gear is selected. Such things point to a hard life and a gearbox that is due for replacement or refurbishment.

The Range Rover braking system is relatively straightforward, with the four discs giving more than ample stopping power. When using the brakes hard to decelerate from speed, watch out for the brakes pulling the vehicle to one side: a sign of worn discs or calipers. Badly bled brakes may result in a spongy brake pedal, which can be cured by re-bleeding the brakes.

INTERIOR

If you intend to buy a Range Rover as your everyday car, a tidy interior is a must. If the interior does need work, however, all is not lost: good second-hand parts are not hard to come by (even full leather interiors can be had at a price) and there are reputable aftermarket companies which provide an excellent rejuvenation service for tired interiors.

However, it is almost impossible to find an early two-door Range Rover with its PVC trim in good condition, and it will be very difficult indeed to find good replacement seats of the right type if you wish to keep the vehicle in original condition. The

Probably the best diesel conversion of all used BMW's turbocharged six-cylinder car engine. Unfortunately, it was too expensive to become popular.

PVC upholstery tends to crack terribly and repair of damaged seats is next to impossible. Fortunately, the other types of fabric used on Range Rover seats over the years are generally very hard wearing, although tears are inevitable on all but the most cherished mid-1970s examples. The pale beige upholstery of early four-doors gets dirty very easily, as does the early silver-grey Vogue type. Generally speaking, if the fabric is not too worn, it should clean up well with a professional valet. If all else fails, aftermarket seat covers will probably prove cheaper than a set of new seats!

Front seats can suffer from saggy bottoms, a fault that is especially prevalent on mid-1970s two-doors. This problem is caused by the rubber seat base diaphragm splitting and can only be cured by replacement. Check for it by looking under the front seat where you can see two rubber rings, a fat one and a thin one. Chafing and general wear in the small ring is common, but damage to the big one will mean that the seat will collapse in due course.

Seat belts are another problem area on the two-door Range Rover. The early static type are frankly nothing but a pain in the neck and the later inertia-reel type are infinitely preferable. These come in two types, fitted either to the seat itself (in late 1970s and early 1980s models) or to the B-pillar (in later vehicles such as the 1990 CSK). If you buy an early vehicle with static belts, it is worth budgeting for a change to inertia-reel belts unless you intend to keep the vehicle strictly original. Remember, though, that these are safety-related items, so take appropriate advice before buying replacement belts.

A final problem with two-door interiors can be found on the lower door trims, which tend to wear easily and distort when the hardboard backing becomes damp. If this had happened, it is likely that the vehicle has seen some fairly serious off-road action, so bear that in mind when you look over the other components and body panels.

Turning to the more luxurious Vogue interiors, the early versions tend to suffer from cracked wood door cappings that can look tatty if not replaced. Make sure that all the electric windows work, as the motors tend to get a little tired with age. Electric windows up to around 1988 generally seem to be slow going up, although as long as the motors still work this should not be too much of a problem.

With later Vogue SE interiors, the leather can crack and age badly if not properly treated: it needs a dose of hide food once in a while if it is to remain supple. The front seat sides are particularly prone to becoming badly cracked and dirty. Ingrained dirt is not particularly easy to remove on leather, especially on lighter coloured types.

One consistent problem that seems to affect all Range Rovers up to the mid-1980s is a saggy headlining. Although the board backing normally stays in place, the cloth itself tends to fall down, resulting in an effect that is known in the trade as 'hammocking'. Hammocking is caused by condensation and cold temperatures affecting the glue, although this is not as serious as the backing board sagging down itself.

The plastic clips which hold the later two-piece headlinings in place across the centre of the vehicle can also fail, thus allowing the headlining to droop where the two pieces meet. Look out, too, for headlining damage above the rear load area; in early vehicles with the one-piece headlining, damage may cause the whole headlining to sag; on all types, it may ruin the appearance of an otherwise tidy interior. Although headlinings are easy to replace (by aftermarket glass-fibre-backed ones if you wish), their purchase price is fairly high. Budget accordingly if the vehicle you plan to buy has a badly damaged one.

SUMMARY

Range Rover prices seem to have remained pretty static over the past few years and in the UK it is still very difficult to buy a decent Range Rover for less than £2,500 (at 1995 prices). There are a lot of tired examples commanding a lot less, but as we have said above, these are only really worth considering as the basis of triallers, or for thorough restoration or rebuild projects.

Four-door models command slightly more than two-doors of the same age, reflecting their greater desirability and versatility. As far as transmissions are concerned, five-speed manuals are preferable to four-speeds, four-speed automatics are preferable to three-speeds, and Borg Warner chain-driven transfer boxes are much quieter than the earlier gear-driven types. Overdrives on four-speed manual models are a definite boon for road use, although they do add to the general level of transmission noise.

The models that offer the best value for money are the mid-1980s four-doors, as these saw the first major improvements in the interior and engine specifications. The fuel-injected 3.5-litre V8 coupled to the five-speed manual or four-speed automatic gearbox is an excellent combination and you might just be lucky enough to find a vehicle that has been cherished from new by its owner and has never been taken off-roading!

In summary, providing you take your time to find the perfect Range Rover to suit you, your budget and your needs, the chosen vehicle should give many years of faithful service. Check it regularly for faults, and make sure that it is regularly serviced. Providing you look after it, depreciation on a Range Rover is minimal and you should be able to get a good price when you yourself come to sell it. After all, there will always be a first time buyer somewhere who wants to experience the joys of Range Rover ownership.

Index